CU00404303

Britain's Game Fishes

Reader feedback

Pelagic Publishing welcomes feedback from readers – please email us at info@pelagicpublishing.com and tell us what you thought about this book. Please include the book title in the subject line of your email.

Publish with *Pelagic Publishing*

Pelagic Publishing publishes scientific books to the highest editorial standards in all life science disciplines, with a particular focus on ecology, conservation and environment.

We produce books that set new benchmarks, share advances in research methods and encourage and inform wildlife investigation for all.

If you are interested in publishing with Pelagic please contact editor@pelagicpublishing.com with a synopsis of your book, a brief history of your previous written work and a statement describing the impact you would like your book to have on readers.

Britain's Game Fishes

Celebration and Conservation of Salmonids

Mark Everard and Paul Knight

Pelagic Publishing | www.pelagicpublishing.com

Published by **Pelagic Publishing**
www.pelagicpublishing.com
PO Box 725, Exeter EX1 9QU

Britain's Game Fishes
Celebration and Conservation of Salmonids

ISBN 978-1-907807-35-0 (Hb)
ISBN 978-1-907807-36-7 (eBook)

Copyright © 2013 Mark Everard and Paul Knight

All rights reserved. No part of this document may be produced, stored in
a retrieval system, or transmitted in any form or by any means, electronic,
mechanical, photocopying, recording or otherwise without prior permission
from the publisher.

While every effort has been made in the preparation of this book to ensure the
accuracy of the information presented, the information contained in this book is
sold without warranty, either express or implied. Neither the authors nor Pelagic
Publishing, its agents and distributors will be held liable for any damage or loss
caused or alleged to be caused directly or indirectly by this book.

British Library Cataloguing in Publication Data
A catalogue record for this book is available from the British Library.
Printed and bound in India by Replika Press Pvt. Ltd.

Cover image is from Trout in the Roots. © David Miller 2002.
www.davidmillerart.co.uk

Contents

Authors' biographies

Dr Mark Everard has a lifelong obsession with fish, water and the aquatic environment. Author of numerous books, magazine and scientific publications, many of them addressing fish and fishing, Mark is also a regular contributor to TV and radio. He is an adviser to government in the UK, India and South Africa on the sustainable use and management of water and other ecosystems, having also advised and conducted research right across the world. Mark is science adviser to the Salmon & Trout Association (S&TA) as well as vice-president of the Institution of Environmental Sciences (IES), fellow of the Linnean Society, founding director of the Bristol Avon Rivers Trust (BART), and a life member and former council member of both the Freshwater Biological Association (FBA) and the Angling Trust. Mark finds time to fish whenever and wherever the opportunity presents itself, but most particularly in rivers accessible from his home in North Wiltshire where he lives with his partner Jackie, daughter Daisy and many tanks of fish.

Paul Knight has been involved with fish for most of his working life, including catching them as a commercial fisherman and angler, growing them as a trout farmer and now trying to conserve them as Chief Executive of the fisheries environmental charity, the Salmon & Trout Association (S&TA). He is a Council member and Fellow of the Institute of Fisheries Management (IFM), and spends his time attempting to influence government departments and agencies to follow policies that will protect the aquatic environment and all its dependent species. He writes for various fisheries-related and environmental magazines, and has successfully published a book of angling tales, *Amazing Fishing Stories*. He lives with his wife, Angela, and son, Archie, in Wiltshire.

Preface

Our world is a rich and marvellous place, home to a bewildering diversity of organisms each elegantly adapted over evolutionary timescales to the environment within which it occurs. All are intimately integrated with the geology, soils and topography, flows of water, chemicals and energy, and the host of other organisms comprising the ecosystems of which they are inter-dependent elements.

However, much of nature remains unknown to us. The endless cycles of substances and energy upon which all living things depend, including all aspects of our own needs – from basic life support to economic activities and aspirations to live fulfilled lives – are reliant upon the ceaseless activities of bacteria and other micro-organisms to a far deeper extent than we are often comfortable to acknow-ledge. We may be familiar with the actual or digital sight of blue whales, roe deer, polar bears, common toads and peregrine falcons. We may, indeed, feel motivated to support charities dedicated to their conservation. This has real value for those conspicuous and charismatic species, and also for the wider, largely invisible ecosystems essential to support them. But let us be under no illusion that what we can see, particularly wildlife we find 'cute' and inspiring, is in reality the tip of the proverbial iceberg of biodiversity upon which all life depends absolutely. Nevertheless, charismatic and economically important organisms have a key role as indicators of the integrity and vitality of the ecosystems that support them, and as a flagship around which public support may be mustered.

Game fishes, particularly members of the salmon family, clearly fall into this category of iconic, charismatic and economically important organisms. They have direct and significant value to anglers and associated tackle and tourism trades, to commercial fishermen and local economies, and may also support traditional livelihoods and regional character. However, their very presence also assures us of a diversity of less direct benefits. For example, thriving populations of salmon, trout and other native fishes send measurable but also subliminal signals to the wider world that the rivers, lakes and seas they inhabit are in a healthy ecological state, as well as being fit to support the diversity of human needs for water and productive, fertile riparian soils. It is not without good reason that the US Environment Protection Agency (EPA) elected to simplify public communication of its often baffling array of water quality standards to the more intuitive strap-line of 'drinkable, swimmable, fishable' (sometimes augmented with 'boatable') to reflect how people intuitively evaluate freshwater bodies. A river fit for its native complement of game fishes, members of the salmon family, as well as other representative species is a river fit for people, able to support our health and other diverse needs into the future.

In this book, we will explore the importance of fishes of the salmon family for the wellbeing of society. There is a largely UK focus to this, but the principles apply wherever in the world game fishes fin through life-giving waters, including all of their life stages whether in fresh or marine environments.

Above all, this book is dedicated to the realisation of rivers, lakes, estuaries and coastal waters fit for the future, serving the wellbeing of all species: fishes, humans and all water-dependent life.

Game for the future; for all; forever.

Dr Mark Everard and Paul Knight

Part 1

The native game fishes of the British Isles

Chapter 1

A natural history of Britain's game fishes

The fishes of the salmon family are many, varied and widespread. Six species are native to British waters, all requiring the highest environmental quality. Four of these we think of as the 'mainstream' game fishes: the Atlantic salmon, the brown trout or sea trout, the Arctic charr, and the grayling. Two less common species are whitefishes, or coregonids, which are not game fishes in the sporting sense of the word. Whilst we will touch upon some details of the natural history of the whitefishes in passing, this book is about the four main game species and so it is to the natural history of these four that we will devote most attention. Then, towards the end of Part 1, we touch upon a 'familiar alien', the rainbow trout, introduced here from its natural range in the Pacific coastal catchments of North America and so in no sense native, but now so widespread in our lakes and rivers that we have at least to give it a brief 'honourable mention'.

1.1 The salmon family

The salmon family, known scientifically as Salmonidae and often referred to as the 'salmonids', is the only living family of the order Salmoniformes within the greater group of ray-finned fishes. Globally, the salmon family includes the salmon, the trout, the charr, the graylings, and the freshwater whitefishes. In evolutionary terms, all of these groups share common ancestors dating back to the middle Eocene era (roughly 37–47 million years ago).

Typically, the fishes of the salmon family are slender, with pelvic fins placed far back on the underside of the body and a conspicuous rayless adipose (or 'fatty') fin towards the rear of the back, behind the single dorsal fin and before the caudal, or tail, fin. The salmonids also possess a fleshy, pointed flap above the pelvic fin base. The scales are rounded and the tail is generally forked, although can become square or even convex in some older trout specimens. The mouths of most of the salmon, trout and charr contain a single row of sharp teeth, whilst the whitefishes

Fig 1.1 The adipose fin of a brown trout.

have only weak teeth, and grayling have none at all. The largest members of the salmon family, including the Atlantic salmon, can reach a body length up to 2 metres (six feet). The smallest member of the family is the pygmy whitefish (*Coregonus coulteri*) found in Lake Superior, at just 13 centimetres (5.1 inches) from snout to tail as an adult.

All fishes are equipped with a host of senses, the relative prominence of which is adapted to their particular life habits. Carp, catfishes, eels and other groups that often thrive in murky waters are liberally equipped with chemical sensory organs across their bodies with which they taste/smell (the sensations merge into a 'general chemical sense') their surroundings, including gradients of chemicals within it. The lateral line system, extending along the flanks of most fish species, including the salmonids, senses changes in water pressure, providing information about currents, obstructions to flow and other creatures moving in the vicinity. Sight is often well developed in fishes adapted to a clear-water, predatory lifestyle, such as the members of the salmon family. However, unlike those of humans and other animals, which are forward facing, the eyes of trout, salmon and many other fishes are on the sides of the head, which equips them with a near-complete field of vision but not the depth perception of our stereo sight. The fishes of the salmon family compensate for this with good colour distinction, better to differentiate detail, but there is also mounting evidence that they can discern polarisation of light, which is invaluable in determining fine details of prey organisms under clear, rippling water.

All members of the salmon family spawn in freshwater. Many stay in fresh-water lakes, rivers and streams throughout their lives. Others are anadromous, meaning that they migrate to sea for a marine adult phase before returning to fresh waters to spawn. These migratory habits, and their ability to run to sea, mean that many members of the salmon family have penetrated many new waters left behind by the retreat of glaciers following the last ice age. Some formerly sea-going populations have subsequently become landlocked, including, for example, the Sebago salmon (*Salmo salar sebago*, which is a subspecies of the Atlantic salmon) in the northern Atlantic USA, and the taimen or Siberian salmon (*Hucho taimen*) of the Caspian Sea region. The Artic charr (*Salvelinus alpinus*) and many of the whitefish species are also commonly landlocked, inhabiting cool, clear freshwater habitats and generally running tributary streams to spawn.

The natural distribution of the salmonid species in the northern hemisphere has been considerably enhanced by their migratory habits and adaptation to cool, clean waters. They are far better at spreading to, and thriving in, these cooler, nutrient-poor waters than other families of fishes widespread and common in the northern hemisphere, such as the carps (family Cyprinidae) in Eurasia and the catastomids (family Catastomidae) in North America. The geographic range of the salmon family is therefore broad across the northern hemisphere, with members found naturally as far south as Spain and Portugal and northwards well into the Arctic circle regions of Norway, Russia, Iceland, Greenland, Canada and Alaska.

In *The History and Topography of Ireland*, Giraldus Cambrensis (or 'Gerald of Wales' c.1146–c.1223), a medieval clergyman and chronicler of his times, stated that, '. . . pike, perch, roach, gardon, gudgeon, minnow, loach, bullheads and verones . . .' were absent from Ireland, also observing that all the Irish species of freshwater fish known to him could live in salt water. Trout, salmon and Arctic charr were prominent amongst the eleven migratory or brackish-tolerant species of freshwater fishes (also including pollan, sticklebacks, eels, smelt, shad, three species of lamprey and the increasingly rare sturgeon) that were able to colonise Ireland's freshwater systems without man's interference. Grayling are notably absent from this list, lacking any sea-going life stage.

All members of the salmon family are predatory, feeding on small animals such as crustaceans, aquatic insects and their larvae, as well as smaller fish. Within these general characteristics, the salmonids exhibit remarkable plasticity of form and lifestyle, exemplified by the wide diversity of brown trout populations that we will consider later in this book. Genetic and behavioural differences, often triggered by environmental conditions, enable these fishes to adapt to different diets, flow conditions, salinity regimes and migratory or residential (sessile) habits. Aside from marine adult life stages, the salmon family is best adapted to a predatory life in clean, fast-flowing streams and rivers, and in cool, clear still waters. Where the members of the salmon family are naturally absent, other groups of fish have evolved to fill this niche of streamlined, fast-water predators. There are many examples of this in Indian rivers, including the riverine Indian trout or trout barb (*Raiamas bola*), the Burmese trout (*Raiamas guttatus*), the barils (various species of the genus *Barilius*) and the copper mahseer (*Neolissochilus hexagonolepis*).

Fig 1.2 Giraldus Cambrensis (Gerald of Wales) documented the fishes of Ireland before his death in 1223.

It is worth noting in passing that, although they possess adipose fins and share a similar streamlined form and predatory, migratory lifestyle, the smelts (including the widespread European estuarine *Osmerus eperlanus*) belong to a closely related group of generally estuary-dwelling fishes, the smelt family (Osmeridae), so are not members of the Salmonidae.

This book is concerned primarily with the four native game species of the British Isles: the Atlantic salmon, the brown or sea trout, the Arctic charr, and the grayling. To each we will devote a short chapter in this first section, then briefly consider the whitefishes as a nod to their 'royal' salmonid blood and note a few details about the rainbow trout, the American interloper.

All of these six native salmonid species are accorded conservation priority, both for inherent value as well as protecting them as exploitable resources, under various conventions and legislation. Amongst these measures is the Bern Convention (Bern Convention on the Conservation of European Wildlife and Natural Habitats 1979), which aims to conserve wild flora and fauna and their natural habitats on a concerted international basis, including both grayling and Atlantic salmon in its Appendix III (species for which exploitation is controlled) and also banning certain destructive means of killing, capture and other forms of exploitation. Some species are also scheduled for protection under the 1992 European Union (EU) Habitats Directive (Council Directive 92/43/EEC on the Conservation of Natural Habitats and of Wild Fauna and Flora) including Atlantic salmon under Annex II (requiring designation of Special Areas of Conservation) as well as grayling, the whitefishes and Atlantic salmon under Annex V (addressing management measures to control exploitation). Fishes of the salmon family scheduled under the UK's Wildlife and Countryside Act 1981 include the two species of whitefish under Schedule 5 (animals strictly protected). In addition, the Atlantic salmon, brown trout, Arctic charr and the two species of whitefish are listed under the UK Biodiversity Action Plan as part of national strategy and supporting action plans to protect or enhance biological diversity.

CHAPTER 2

The Atlantic salmon and
its amazing life-cycle

T he Atlantic salmon (*Salmo salar* Linnaeus, 1758) is well known as an iconic
fish of European rivers, as a popular food and sporting species, and for
its extraordinary life-cycle. Famously, and often repeated, Izaak Walton's
'Piscator' notes that, 'The Salmon is accounted the King of freshwater fish . . .'.
Aside from a few landlocked subspecies, Atlantic salmon are anadromous, migratory
fish running to sea to feed into their adult form but returning to fresh waters to
spawn, generally but not exclusively in their natal rivers.

2.1 Key features of the Atlantic salmon

Atlantic salmon naturally occur in the temperate and Arctic regions of the
Northern Hemisphere bordering the Atlantic Ocean, as well as the Baltic and the
Barents Sea north of Russia. The west of this range includes coastal drainages
from northern Quebec in Canada and Connecticut in the USA. There are some
Atlantic salmon populations in Argentina, but these were introduced by man. To
the east, Atlantic salmon are found in drainages from the Baltic states to Portugal.
Furthermore, landlocked stocks of Atlantic salmon occur in Russia, Finland,
Sweden and Norway and, in the case of the Sebago salmon, in North America.

The dorsal fin of an Atlantic salmon has 3–4 fused spines supporting between
9 and 15 soft rays, whilst the anal fin has 3–4 spines supporting 7–11, soft rays.
There are 58–61 vertebrae in the backbone of the fish, supporting a streamlined
body evolved to cope with fast swimming and strong water flows. Adult fish
have a blue-green body colour overlain with a silvery guanine coating, generally
with few spots in their marine life phase, none of which occur below the lateral
line.

2.2 The life-cycle of the Atlantic salmon

Salmon eggs hatch during early spring in depressions known as 'redds'. These are excavated by female fish during the previous winter's spawning in the well-flushed gravels of cool and clean upper reaches of rivers. These hatchlings are known as alevins, characterised by a yolk sac on which they feed for a number of weeks whilst remaining in the protection of the redd. Once the yolk is consumed, the young fish emerge from the gravel in April or May as fry, typically with a body length of 2.5 centimetres (one inch). The fry feed on small river invertebrates, growing into parr, which develop between 8 and 12 characteristic dark blue-violet 'dirty thumbprint' markings, known as parr marks, along their flanks.

Parr are voracious feeders and, although often gregarious and found in considerable numbers occupying suitable gravel riffles and runs in stronger-flowing rivers, are territorial, vying for good feeding locations amongst their peers. Populations of salmon parr in rivers are therefore 'density-dependent', limited by both the availability of a suitable riffle habitat and the territorial space taken by each parr. The availability of a suitable habitat is therefore an important limiting factor to salmon populations in this early life phase.

The duration of the parr life stage is highly variable across the natural range of the Atlantic salmon, dependent upon environmental variables such as water temperature, habitat suitability and the availability of food. During this time, the young fish are vulnerable to a wide range of predators, including birds such as herons, fish such as pike, and mammals including otters and mink. However, after typically one to three years, but as much as six in some cold, upland streams, the parr begin the process of smoltification in which they metamorphose into smolts. This process involves the juvenile fish improving their hypo-osmoregulatory performance (their capacity to pump salts across the gill interface to excrete excessive concentrations from the blood into surrounding sea water) and developing a flush of silver scales. Once fully metamorphosed, smolts then have the ability to occupy saline waters, and they move downstream to exit river systems for a marine adult life phase.

Generally, smolts remain in the vicinity of the estuary for some time, perhaps whilst undergoing full adjustment to higher salinity water, and are therefore extraordinarily vulnerable to pollution or poor water quality, excessive predation or

Fig 2.1 Salmon parr with characteristic 'inky fingerprint' markings.

Fig 2.2 Disko Bay, Greenland, where salmon come to feed.

fishery activities, and other stresses such as diseases and parasites arising from fish farms. As they head off to open-sea feeding areas, smolts demonstrate schooling behaviour. Evidence from the USA suggests that recently spawned adults (kelts) returning to the sea may act as guides to the juveniles at this stage.

As marine predators, smolts rapidly mature into adult salmon, feeding voraciously on a variety of krill and shrimps, squid, other invertebrate prey and small fish. Their growth is rapid on a far richer larder of food than was available in their natal rivers. Bigger salmon feed largely on other fishes such as herring, alewives, smelts, capelin, small mackerel, sand launce (sand eels), blue whiting and small cod. Populations of Atlantic salmon breeding in British waters occupy a range of marine areas, of which the best known are in the Norwegian Sea and the fertile waters of the continental plate off south-west Greenland.

Salmon typically begin their spawning migration back into fresh waters after between one and four years at sea. The triggers for this migration urge are a mix of behavioural and genetic, but the scent of home rivers is a powerful signal. Salmon that remain at sea for more than one winter undertake the longest migrations, whilst one sea winter fish, known as grilse, tend not to travel beyond the Faeroe Islands and the southern Norwegian Sea. In this way, individuals from one year class of salmon spend differing lengths of time at sea, and also occupy

Fig 2.3 A. F. Lydon's painting of a grilse from the Rev. W. Houghton's *British Fresh-water Fishes* (1879).

different areas, so adding security to the species by safeguarding against catastrophic events affecting any one year class in either their marine or freshwater phases.

Once they re-enter river systems, salmon cease to feed, living on reserves of fat stored in their tissues for as much as a year before they eventually spawn in the gravel-bedded headwaters in which they hatched. A very few fish will stray to other rivers and breed within different populations, so allowing a regular ingress of new blood.

Mature Atlantic salmon may lie quiescent in river systems for some considerable time. Adult fish returning to British rivers in late winter or early spring will stay in freshwater until spawning begins in the Autumn, while some Russian salmon enter rivers on the Kola Peninsula in September of the year preceding that in which they will eventually breed.

During this time, they change significantly in appearance. Male fish develop a powerful kype, or hooked lower jaw, which they use as a weapon to defend spawning sites later in the year. They also grow progressively thinner as they consume internal fat reserves. Atlantic salmon also change considerably in body colour, losing their silvery guanine coat and becoming greenish or reddish brown, mottled with red or orange. This colour change is brought about by a remobilisation of carotenoid pigments, derived from the invertebrates upon which they fed at sea and which were laid down in body fat. The colour change is particularly prominent in cock (male) fish, the pigments within hen (female) fish being remobilised with fat reserves into the production of yolk in their maturing eggs. This affects not only the appearance but also the taste of the flesh, and therefore most salmon caught commercially for food are intercepted at sea, in estuaries or at the bottom of river systems, where they are in their most pristine condition.

The heroic efforts of mature salmon to pass obstructions in fast-flowing water are well known, adding to their iconic status and constituting one of nature's most spectacular sights: a staple of many a television documentary. Izaak Walton's description from his classic 1653 book *The Compleat Angler* has rarely been bettered:

. . . they will force themselves through floodgates, or over weirs, or hedges, or stops in the water, even to a height beyond common belief. Gesner speaks of such places as are known to be above eight feet high above water. And our

Fig 2.4 Mature salmon resting before attempting to pass through a weir.

Camden mentions, in his *Britannia*, the like wonder to be in Pembroke-shire, where the river Tivy falls into the sea; and that the fall is so down-right, and so high, that the people stand and wonder at the strength and sleight by which they see the Salmon use to get out of the sea into the said river; and the manner and height of the place is so notable, that it is known, far, by the name of the Salmon-leap.

The timing of spawning varies between rivers, and may also be significantly influenced by water temperature, the length of daylight (which influences hormonal triggers), as well as surges of water (spates) down rivers, encouraging and enabling migratory runs through, and over, obstacles. In Great Britain, salmon generally spawn between November and December. Male fish vie with each other for the best, well-flushed coarse spawning gravels, holding them as territories from which they drive away competing cock fish as well as other species. Here, they entice ripe hen salmon to excavate 'redds' in the gravel bed using powerful body movements, into which eggs are then deposited and milt shed by the cocks to fertilise them. Even at this stage, nature does all in her power to add to the security of the species, by enabling 'precocious parr', which have matured sexually though have not yet completed their transformation into sea-going adult fish, to join the adults in their breeding frenzy. The eggs from a female fish can therefore be fertilised by an adult cock fish and several parr from a completely different year class, thus providing the maximum possible opportunity for successful spawning and the maintenance of genetic diversity.

Salmon eggs are heavier than water, and so they drop down into cavities in the coarse gravel redds. Many eggs are lost to predators such as invertebrates and smaller fishes, including salmon parr from previous generations. Female salmonids, including the Atlantic salmon, generally lay about 1,750 eggs per kilogramme body weight, so a large female salmon of 20 pounds (9.1 kilograms) may lay as many as 16,000 eggs in one or more redds. Once spawning has been completed, the hen fish then covers the redd using her body to mobilise gravels from upstream. Sometimes, new redds are created in gravels used to cover egg-laden redds, into which the hen fish may deposit further eggs until she is fully spent.

Fig 2.5 Salmon redd.

The eggs lie dormant in the gravel through the colder months, flushed by oxygen-rich river flows until triggered into hatching in the warmer days of early spring. The exact timing of development is related directly to temperature. As ectotherms (cold-blooded animals), the body temperature of which fluctuates with the external environment, the metabolism of salmon and their eggs varies with temperature. As such, egg and fry development and growth is directly related to 'degree-days' (mean temperature in degrees centigrade multiplied by number of days). Eggs of the Atlantic salmon take approximately 220–250 degree days to 'eye-up' (the stage at which a distinct eye is visible through the egg shell), and a further 220–250 degree days to hatch. After hatching, the alevins live off their yolk sacs for approximately 300 degree days, hiding within the gravel until all of the yolk is consumed, after which they leave the redd to begin to feed as free-swimming fry.

The act of spawning and the preparatory run up rivers, exerts a heavy toll on the starving mature fish. These empty, spent fish are known as kelts. Kelts can usually be identified by the thin shape, distended vent and presence of 'gill maggots' (parasitic flukes) on their gill filaments. Immediately after spawning, kelts are weak and many may carry injuries from territorial fights, attack by would-be predators or from passing obstructions on their route upriver. All of these factors make kelts highly susceptible to secondary infections, particularly from the fungus *Saprolegnia*. Many Atlantic salmon, perhaps as many as 90–95% across Great Britain, die following their first spawning. Given the dramatic change in body form and the exertions of fighting other fish for the best river reaches for redd-digging, spawning is almost always fatal for male fish. However, some

Fig 2.6 A dramatic depiction by renowned wildlife artist Robin Armstrong of a mature Atlantic salmon leaping a waterfall.

survivors, predominantly female, regain condition. Some are encountered in rivers by anglers in spring, when they have regained a silvery appearance, and these may be mistaken for fresh run 'spring' fish. Those salmon that survive the rigours of spawning return to sea to feed and rebuild condition before the next spawning run. This stressful existence does not suit the Atlantic salmon to a long life, although the oldest recorded fish reached thirteen years, during which it made several migrations back into freshwater to breed.

The life-cycle is flexible, adapting to local conditions with many variants. These include the Sebago salmon and other landlocked populations noted previously, as well as stocks that breed primarily after one sea winter and others that spend longer at sea to maximise the strength and fecundity of brood fish returning to spawn in particularly torrential rivers. Such is their variability, including changing body forms throughout the life-cycle and the development of discrete genetic strains of fish best suited to particular rivers, that the Atlantic salmon has, in times gone by, been classified into a range of 'species'. However, just the single species of Atlantic salmon, *Salmo salar*, is recognised by scientists today.

2.3 Survival in adverse conditions

Atlantic salmon are distributed southwards to a latitude of about 40 °N to both the eastern and the western shores of the Atlantic Ocean. However, this distribution has been significantly influenced by geological and climatic changes over millennia, particularly including the ice ages.

In France, there is a predominantly autumn run of grilse in Brittany, where spring salmon are today something of a rarity. (There is also a significant sea trout run in Normandy.) To the south west of France, the mild climate and rich flood-plains are fed by cool rivers draining the Pyrenees, including the Gave d'Oloron, which is considered by many to be France's finest salmon river. On the Atlantic seaboard of northern Spain, some twenty-six rivers draining the Cantabrian range and the Galician Coast are run by Atlantic salmon, where fly fishing is popular and salmon feature in local festivals and folklore. Atlantic salmon also run two

Fig 2.7 Fresh grilse and stale 'springer' salmon, different strategies for survival and reproduction that spread the risks in changeable environments.

rivers in northern Portugal, marking their southernmost extent on the eastern shore of the Atlantic Ocean. These two Portuguese rivers, the Rio Minho and the Rio Lima, both rise in higher ground in northern Spain but soon cross the border with Portugal.

These Iberian and southern French populations may have been of great significance for the survival of salmon during the last ice age some 10,000 years ago. Genetic studies suggest that current northern European salmon populations derive from just two distinct stocks, one believed to have survived extensive European ice cover in a south-eastern ice lake refuge and the other occupying southern rivers remaining free from the advancing ice sheet.

A similar refuge strategy during the last ice age is believed to have occurred in North America. Although there remains considerable uncertainty, natural populations of Atlantic salmon in the United States probably date from the end of the last ice age. After this time, Atlantic salmon were probably present in all watersheds from Labrador in Canada southwards to the Hudson River in New York State in the USA. However, wild Atlantic salmon populations in the USA today occur only in Maine, whilst Canadian stocks are generally healthier. Modern American and Canadian stocks of Atlantic salmon take part in extensive marine migrations, particularly to the waters off western Greenland where they merge with European stocks to form a large mixed-stock complex. It is probable that this behaviour enabled Atlantic salmon to re-colonise the rivers of North America after the last ice age, thanks to the 'straying' behaviour in a proportion of mature salmon. This has enabled them to colonise new waters beyond their native streams.

A similar pattern has been observed for the various Pacific species of salmon, which took refuge from the last ice age in areas such as the southern Oregon coastline, the California coastline and the Queen Charlotte Islands of British Columbia, then reinvaded rivers to the north as ice sheets retreated.

Although today we tend to focus on their vulnerability to the multiple environmental pressures that human activities place upon them, if we take a longer-term view then it becomes apparent that the Atlantic salmon is an aggressively invasive species elegantly adapted to riding out climatic adversity and exploiting new habitats as environmental conditions change. From the 'straying' of a proportion

Fig 2.8 A. F. Lydon's painting of an Atlantic salmon parr from the Rev. W. Houghton's *British Fresh-water Fishes* (1879).

of mature salmon beyond their natal streams that enables them to colonise new river systems and perhaps even continents, through to the 'insurance policy' strategy of single and multiple sea winter fishes as well as precocious parr, the potential to complete the life-cycle in landlocked populations, which may perhaps even ride out ice ages in ice lake refuges, and entry into freshwater more than a year in advance of spawning by some Russian salmon in the extreme north, the species has evolved to endure climatic extremes and aggressively to exploit the niches that they leave behind.

2.4 The value of Atlantic salmon stocks

Commercial fisheries for Atlantic salmon are of substantial value across their geographic range. This will be addressed in greater detail in Part 2 of this book, which also addresses salmon aquaculture, a substantial industry in the British Isles as well as around the world, including beyond the natural range of the species.

Izaak Walton's quote about salmon as the 'King of the fishes' is related largely to its angling virtues, which have been enjoyed by the upper echelons of British society, but also increasingly by other sectors, throughout the past two centuries. Recreational Atlantic salmon angling is the basis of a substantial tourism trade across Scotland, Wales and the south west and north of England, as well as supporting a variety of local traditional livelihoods. Salmon can be fished for by various means, although many anglers consider that the 'take' from a salmon on the fly is the ultimate game fishing experience. It is certainly an electric moment when that long pull on the line finally comes; the exhilarating experience of such a first take stays with many an angler for their whole life. However, salmon angling is incredibly water dependent. Too little height and pace, and the fish will not run, or those that already have entered freshwater remain dormant. Too much water, and fly fishing becomes impractical. It is then that spinning comes into its own, and the artificial spinner, spoon or plug can be the most deadly of catchers, as can bait fishing with worm or prawn where permitted.

Once hooked, salmon respond in a variety of ways. Grilse will often rip line off the reel and go on long, exciting runs, as though trying to return to the safety of the sea. Larger fish tend to bore deep and use their incredible power to try and

Fig 2.9 Fly fishing for salmon, considered by many to be the pinnacle of angling experiences.

shake the hook, although when they, too, decide that the downstream option is preferable, it is often a mighty battle to try to keep them in a pool against the flow of the river. Many a fisherman has received a heart-stopping soaking from having to follow a big salmon down through rapids and lower pools before finally subduing the fish or, as often as not, losing it! Whatever the outcome, salmon fishing provides some of the most exhilarating angling experiences possible.

The British rod-caught record for the Atlantic salmon shares the distinction of not only being one of the oldest, but also one that is held by a lady angler. This fish weighed a phenomenal 64 pounds (29 kilograms) and was taken by Miss Georgina W. Ballantine of Glendelvine, whilst fishing the River Tay way back on Saturday 7th October 1922. Published accounts vary in some details, but concur that Miss Ballantine was born, lived and died in a cottage just a decent cast away from the Boat Pool where the fish was hooked. She was fishing with her father, a ghillie, who was in charge of the oars of the boat from which Miss Ballantine began fishing. She was reportedly spinning a 'dace', or in other accounts a 'lacquered dace' or 'spinning dace minnow bait', which may have been an artificial lure or a preserved deadbait on a Malloch's spinning mount. The mighty salmon grabbed the bait at around a quarter past six in the afternoon as it was getting dark and the pair were preparing to leave the river. The fish initially moved slowly downstream, the pair beaching the boat for Miss Ballantine to fight it from the bank. However, they quickly took to the boat once again as 'the Beast', as Miss Ballantine had come to call it, made a run down the river and through the near-side support of Murthly Bridge, out of the Boat Pool. After a dogged fight of two hours and five minutes, the fish was eventually landed at twenty past eight in the evening over half a mile down the river. The rest, as we now know, is history!

2.5 The role of Atlantic salmon in aquatic ecosystems

Atlantic salmon are intimately adapted to the ecosystems with which they coevolved. As predators, prey, hosts to parasites, conveyors of the productivity of the seas into nutrient-poor headwaters, and in many other ways, these fish play

Fig 2.10 A. F. Lydon's painting of a male salmon from the Rev. W. Houghton's *British Fresh-water Fishes* (1879).

important roles in food webs and the functioning of wider ecosystems. They also make critical contributions to the life-cycles of other organisms.

One of the most complex and dramatic of such interrelationships is with the pearl mussel (*Margaritifera margaritifera*), a severely endangered mollusc found living in the bed of just a few exceptionally clean English, Welsh and Scottish rivers. Pearl mussels filter-feed from river water. Between June and July, male mussels discharge their sperm into river water, which is inhaled by female mussels through their filter feeding activities. Inside the mantle cavity of the female mussel, the sperm then fertilise eggs in special sacs adjacent to the gills. From these fertilised eggs, glochidia larvae are released into the water column in great densities, a single female pearl mussel releasing as many as 4 million larvae.

The next stage of the life-cycle depends upon the glochidia finding an Atlantic salmon or brown trout host. Once located, or more likely breathed in by chance, the hinged body of the glochidia larva snaps shut on the gills of the fish where it forms a cyst and continues to grow and metamorphose. Pearl mussel larvae remain attached to the host fish until the following spring or early summer. At this point, they drop off as seed mussel, requiring clean, fine gravel in which to burrow to begin their often long adult lives. Pearl mussels grow only slowly, maturing sexually at between ten and fifteen years, though they can live as long as 120 years.

The critically endangered status of the pearl mussel appears to be due to multiple pressures, significantly including declining water quality and sedimentation, but their dependence upon strong populations of salmonid fishes is also an essential link in the successful completion of their remarkable life-cycle.

The threats facing Atlantic salmon have given rise to a range of conservation measures recognising both their inherent value and the need to protect them as exploitable resources. Details of these protocols and legal instruments are included in the previous chapter. Exploitation of Atlantic salmon, along with grayling, is listed under Appendix III of the Bern Convention, which also prohibits certain destructive means of killing, capture and other forms of exploitation. The Atlantic salmon is also scheduled under Annex II (requiring designation of Special Areas of Conservation) and Annex V (requiring management measures to control exploitation) of the EU Habitats Directive. In addition, the Atlantic salmon is

Fig 2.11 Salmon have often featured in woodcuts and other artwork.

prioritised nationally for protection or enhancement of populations under the UK Biodiversity Action Plan, and various Salmon Action Plans are in place across the UK.

The Atlantic salmon is truly an environmental icon. The incredible life-cycle of this fish, the length of its oceanic migrations and its varied and widespread range make this species perhaps the most iconic of all the fishes. As such, they are also amongst the most managed and protected of aquatic species, as we shall explore further in Part 2 of this book.

CHAPTER 3

Brown trout or sea trout

T he brown trout, or sea trout, goes by the scientific name of *Salmo trutta trutta* Linnaeus, 1758. However, as we will see, it has been known by many other names in the past. The remarkable plasticity of form and lifestyle exhibited by members of the salmon family is exemplified by the many different forms of the brown or sea trout and their associated life-cycles. Indeed, so diverse are the habits and physical appearances of brown trout in different habitats that many have formerly been considered as separate species.

For example, in *The Compleat Angler*, Izaak Walton's 'Piscator' observed that:

> There is also in Northumberland a Trout called a Bull-trout, of a much greater length and bigness than any in these southern parts; and there are, in many rivers that relate to the sea, Salmon-trouts, as much different from others, both in shape and in their spots, as we see sheep in some countries differ one from another in their shape and bigness, and in the fineness of the wool: and, certainly, as some pastures breed larger sheep; so do some rivers, by reason of the ground over which they run, breed larger Trouts.

English, Welsh, Scottish and Irish rivers were colonised by sea trout at the end of the last ice age, and their descendents are the populations of brown trout and sea trout we know today. Resident and migratory characteristics have developed within individual catchments, so that some fish now remain permanently resident (brown trout), some always migrate (sea trout) and others can mix the two life histories, depending on circumstances. It is believed that both genetics and environmental issues, such as habitat and food availability, play a part in whether or not a trout migrates to sea. This would account for instances where sea trout smolts continue to emanate from resident brown trout populations above man-made impassable barriers, such as the Kielder dam in Northumberland.

The sea trout has many local names, including 'sewen' (the silver one) in Wales, 'peel' in the south west of England, and 'finnock' (young fish) in Scotland. They are also sometimes known as 'salmon trout' in culinary circles.

Fig 3.1 Head of a sea trout, salmon-like in appearance.

3.1 Key features of the brown or sea trout

The brown or sea trout is found in clean and cool streams, rivers and lakes, with the sea trout also occupying estuaries and coastal seas. They are widespread on the north western coast of Europe, southwards through Biscay, across North Africa and central European lakes, and into the north west coast of Asia. They have also been widely introduced by man elsewhere in the world.

This trout shares the same streamlined body form as Atlantic salmon, with similar numbers of fin spines and rays (3–4 spines and 10–15 soft rays in the dorsal fin with 3–4 spines and 9–14 soft rays in the anal fin), with a small head and a large terminal mouth armed with short, strong teeth, and extending back behind the eye. Body colouration varies markedly with the habitat and life-cycle of the specific trout population, but is generally a grey-blue background coloured with numerous spots, which occur both above and below the lateral line. Many fresh-water populations have the buttery-brownish background body colour that gives the brown trout its name, whilst sea-going fish go through the same smoltifying process as salmon, producing a similar silver guanine sheen. However, brown trout smolts tend to be bigger than Atlantic salmon smolts.

The amazing diversity of forms of brown trout have been classified as different species over the years, as the introductory quote from Walton indicates. Walton's 'Piscator' also seems to classify sea trout separately from brown trout, describing it thus in *The Compleat Angler*:

> There is also in Kent, near to Canterbury, a Trout called there a Fordidge Trout, a Trout that bears the name of the town where it is usually caught, that is accounted the rarest of fish; many of them near the bigness of a Salmon, but known by their different colour . . . And so much for these Fordidge Trouts, which never afford an angler sport, but either live their time of being in the fresh water, by their meat formerly gotten in the sea. . . .

Fig 3.2 A. F. Lydon's painting of a Gillaroo trout from the Rev. W. Houghton's *British Fresh-water Fishes* (1879).

Fig 3.3 A. F. Lydon's painting of a Lochleven trout from the Rev. W. Houghton's *British Fresh-water Fishes* (1879).

Echoing the received understanding of the age, the Reverend W. Houghton, in his classic and beautifully-illustrated 1879 book, *British Fresh-water Fishes*, lists not one but twelve species, including the salmon trout, sewen, bull trout, Galway sea trout, short-headed salmon, silvery salmon, common trout, black-finned trout, Loch Stenness trout, Lochleven trout, gillaroo trout and great lake trout. All are, of course, classified today as variants of just the single but incredibly variable species *Salmo trutta*. Some of the more extreme forms of the brown trout include the anadromous sea trout, the Irish gilleroo, which has evolved to feed almost entirely on freshwater crustaceans, and the giant predatory lake ferox.

Ferox live longer than other freshwater forms of brown trout, growing larger and developing fearsome jaws consistent with their largely piscivorous diet in deep, cool lakes. Here, they eat prey fish (generally shoals of charr but with powan substituting this role in Loch Lomond) rather than insect life, and they also grow far larger, reaching in excess of twenty pounds in weight. Ferox are a prized angling quarry for those brave and dedicated enough to endure conditions in the central European glacial lakes, the Scottish lochs and the Irish loughs in which they occur.

The extension of the mouth behind the eye, the presence of spots on the body below the lateral line and the lack of the pronounced 'wrist' before the tail fin are

Fig 3.4 A. F. Lydon's painting of a Bull trout from the Rev. W. Houghton's *British Fresh-water Fishes* (1879).

features often used by anglers to distinguish smaller salmon from larger sea trout. However, hybrids between the two species are far from uncommon (estimated at around 7% of the British population) due to similar spawning habits.

3.2 The life-cycle of the brown or sea trout

Brown trout always breed in fresh water, preferring cool, well-oxygenated upland rivers, which provide suitable spawning gravels. Freshwater resident stock generally

Fig 3.5 A trout redd cut in Suffolk's River Lark (with thanks to Andy Taylor, the Wild Trout Trust).

Fig 3.6 A. F. Lydon's painting of a brown trout parr from the Rev. W. Houghton's *British Fresh-water Fishes* (1879).

mature at between three and four years, running up rivers and out of lakes into tributary streams to breed. As with salmon, male trout guard territories within which hen fish cut one or more redds in coarse gravel, laying as many as 10,000 eggs (following the same 'rule of thumb' as salmon and many other salmonids of depositing about 1,750 eggs per kilogramme body weight).

Although sea trout have a similar freshwater biology to salmon, their marine stage is different. They spend between one and five years in fresh water, depending on environmental conditions, and anything between six months and five years at sea. However, they do not normally undertake oceanic migrations as far afield as salmon, preferring to remain in coastal waters to feed on crustaceans and small fish such as sand eels. This makes them more vulnerable to human interference, such as pollution, parasite transfer from marine cage aquaculture units, and commercial fishing. However, we still lack sufficient knowledge about this stage of the sea trout's life to be certain about which management regimes afford adequate stock protection. Also like Atlantic salmon, sea trout run rivers to spawn in their natal headwaters, where their spawning behaviour is similar.

Trout eggs hatch in the well-flushed gravels of redds in late winter or early spring, after some 500 degree days' incubation. The alevins then live off their yolk sacs for a matter of weeks before emerging as fry to feed on small invertebrates. As the fry mature into parr, and then into adult brown trout or smolts, they feed mainly on aquatic and terrestrial insects as well as molluscs, crustaceans and small fish. Trout can be long-lived in some environments, particularly ferox, reaching a maximum reported age of 38 years.

3.3 The value of brown or sea trout stocks

Whereas the majority of trout for the table, including both sea trout and brown trout, were taken from the wild up until the 1950s, commercial trout aquaculture (which we will discuss in more detail in Part 2 of this book) has since substantially displaced the wild harvest of non-migratory trout.

Nevertheless, sea trout are still largely harvested from the wild with traditional methods, such as Welsh net fishermen, given additional protection in by-laws updated in early 2008 by the Welsh Assembly Government and Environment Agency Wales. The purpose of these by-laws is to protect the future of the traditional net

Fig 3.7 B. Fawcett's engraving 'On the Wye'.

fishermen and local aspects of the Welsh economy, but also the salmon and sewen (sea trout) that they and the recreational angling community and economy rely upon. Various regionally distinct forms of net fishing have taken place on Welsh rivers for centuries, ranging from seine net fishing in the Dyfi, Dysynni, Conwy, Mawddach, Nyfer (Nevern), Teifi and Tywi, compass nets on the Daucleddau, lave nets on the Wye, ancient basket traps on the Conwy, and coracle fishing in Carmarthenshire and Ceredigion.

Measures in the 2008 Welsh byelaws (applied to England as well) include banning of the sale of rod-caught salmon and sewen, and the introduction of mandatory carcass tagging of net-caught fish to prohibit sale of illegally caught fish. Each of these green tags is highly visible and bears a unique code, which can be traced directly back to the individual licensed net fisherman. Buyers are now able to identify legal and illegal fish, addressing directly the market for the criminal activity of poaching, which is a significant threat as sea trout become vulnerable to netting on entering smaller streams and pools. This should make enforcement much easier, as effort can now be much more efficiently placed on market inspections rather than trying to catch poachers red-handed generally at night out on the river or estuary. Added to these netting byelaws, the fishing season and the number of licenses granted are both controlled to limit pressure on already threatened sea-trout stocks.

Recreational angling for brown and migratory sea trout is a major industry. Brown-trout fisheries range from stocked commercial pools, to lakes, reservoirs and rivers, which may range in size from lowland reaches to tiny hill streams. Concerns have rightly been raised about the volume of farmed brown and other trout being stocked across the British Isles, not merely from the point of view of competition with wild fish but also that fertile strains may dilute or suppress the genetic diversity of native populations. Environment Agency guidance in its 2008 document, *Better Sea Trout and Salmon Fisheries: Our Strategy for 2008–2021*, recommends a progressive transition either to stocking with native provenance trout or else with increasing proportions of sterile (generally triploid) brown trout to avoid genetic impacts upon native stocks. Natural England, the statutory wildlife regulator for England, also campaigns for only native provenance fish to be introduced into waters. Controls on pressures likely to perturb the natural ecosystems, including fish stocks, of water bodies is also required in European legislation. The future of stocked brown trout is likely to be changeable, though the scale of interest in recreational angling for the species remains high and supports substantial economic interests, including angling tourism, tackle, hospitality and related trades.

However, there is a premium on wild, self-sustaining stocks of brown trout, which clearly relies not on stocking and management of rivers and pools for access but rather restoration of river habitats to optimise natural self-sustaining stocks. As such, the strong and growing interest and economic opportunity around wild fish can be a significant force for good in terms of the protection and rehabilitation of river ecosystems. This benefits not merely trout and anglers, but also wider biodiversity and nature conservation, associated ecotourism, improved

Fig 3.8 Fly fishing for brown and sea trout is a rewarding pastime.

Fig 3.9 Trout too have often featured in woodcuts and other artwork.

Fig 3.10 Sea trout.

quality and quantity of fresh waters flowing down to users in the lowlands, protected landscapes and a host of additional benefits besides.

Angling for sea trout has a mystique all of its own, undertaken mainly by casting a wet fly by night as the sea trout become active. It is a sport for the passionate, but attracts significant devotees, and therefore represents a famous tradition as well as supporting regional economies around sea-trout rivers. This is especially important in systems where salmon runs have declined over recent years and sea trout have, in many cases, become the priority quarry for migratory game-fish anglers. In England and Wales, the Environment Agency appreciated this change in emphasis when it published its new strategy for migratory fish management in 2008. No longer did this concern itself just with Atlantic salmon; the new strategy was titled *Better Sea Trout and Salmon Fisheries*. The once 'forgotten' migratory fish could be said to have been brought in from the cold.

This situation has been further strengthened by a plethora of scientific projects aimed at researching what happens to sea trout in their marine phase, including stocks in the North Sea, English Channel, Wales and Ireland.

3.4 The role of brown or sea trout in aquatic ecosystems

Along with Atlantic salmon, native trout can also play host to pearl mussels, as well as other species of freshwater mussel, not to mention a complement of other parasites. Also, perfectly adapted as they are to life in faster flows, streamlined in body form and able to make use of the slightest lees in rocks and other obstructions to hold station as they wait for food items to drift past or above them, brown trout are important parts of the food chain for upland aquatic ecosystems. Sea trout, too, serve this role on their return, also bringing with them marine nutrients stored away in their bodies and eggs as they run rivers to spawn.

The conservation of wild stocks of brown and sea trout is recognised as a priority for both their inherent value as well as for sustainable exploitation. This

Fig 3.11 A.F. Lydon's painting of a 'sewen' from the Rev. W. Houghton's *British Fresh-water Fishes* (1879).

Fig 3.12 Wild brown trout taken on a fly from an untamed stream . . . bliss!

includes under the UK Biodiversity Action Plan as well as the Environment Agency's 2008 strategy *Better Sea Trout and Salmon Fisheries*, both of which have given rise to various local action plans as well as increased controls on stocking with fertile farm-reared trout likely to dilute the genetic integrity of localised native trout populations.

A babbling stream without a brown trout is incomplete, demonstrating to anglers and the wider public alike that something may be amiss. They are, therefore, something of an icon of a vibrant river or clean lake environment, simultaneously of nature conservation, commercial and broader societal value.

Image from the Dr Robert Holland Collection, courtesy of the Freshwater Biological Association

CHAPTER 4

The Arctic charr

A rctic charr (sometimes spelled Arctic 'char'), *Salvelinus alpinus alpinus* (Linnaeus, 1758), is both a freshwater and saltwater member of the salmon family, native to the Arctic and Subarctic as well as alpine lakes and coastal waters. It is believed that no other freshwater fish is found as far north, not even the three-spined stickleback (*Gasterosteus aculeatus*), which is also particularly hardy. Deriving its name from its preferred environment, the Arctic charr is the most common and widespread salmonid fish in Iceland, and is the only species of fish in the far-northern Lake Hazen, the northernmost lake in Canada on Ellesmere Island in the Canadian Arctic.

4.1 Key features of the Arctic charr

The Arctic charr is closely related to both salmon and trout, and shares many of their characteristics. The fish has an elongated and streamlined oblong body, rounded in cross section, and a small head and large terminal mouth, armed with teeth for a mobile, predatory lifestyle. The scales are very fine and also deeply embedded, giving the skin a smooth, slippery feel. Also, unlike trout, charr possess teeth only in the central forward part of the mouth. Other differences between species of charr and trout are less obvious, but include a distinctive boat-shaped bone in the upper part of the mouth of the charrs. Body colour varies dramatically with habitat. It is, indeed, so variable in fresh waters that, much as for the brown trout, different strains were once thought of as separate species, though there is a predominance of greens, reds and ambers. (This leads some anglers to refer to Arctic charr by the nickname 'old tartan-fins'.)

The anadromous form of the Arctic charr, which makes annual migrations to sea, tends to grow far larger on the more plentiful food resources in the ocean, and also loses its characteristic markings to take on an overall silvery sheen with a dark back more akin to that of salmon or sea trout. However, as the mature fish return from the sea to spawn, the belly becomes red and the sides are brownish, often with an intense yellowish-green tinge. Landlocked Arctic charr generally

Fig 4.1 A. F. Lydon's painting of a Windermere charr, Cole's charr and Gray's charr from the Rev. W. Houghton's *British Fresh-water Fishes* (1879).

occur where they are blocked from the sea by some physical barrier, and are found everywhere in the northern sea-run charr's territories, but also occur in smaller numbers further to the south. The charr is generally considered a glacial relict, stranded in cold, deep lakes as far south as New England, Switzerland and Great Britain.

Like brown trout, charr also exhibit a great diversity of lifestyles, prompting nineteenth-century scientists to be as inventive in the identification of subspecies as for the brown trout. In all, there were once considered to be some 31 separate subspecies of charr, based on often minor idiosyncrasies amongst isolated populations induced by specific sets of light, temperature, food and other local conditions. In his 1879 book, *British Fresh-water Fishes*, the Reverend W. Houghton listed six British 'species' of charr: Windermere, Cole's, Gray's, Alpine and Loch Killin charr and also the torgoch (or Welsh charr). Today, however, modern science recognises only the single species, *Salvelinus alpinus*, in the British Isles; 'torgoch' remains the local name for Arctic charr in the lakes of Snowdonia in north Wales. (Four types of charr are recognised as true species in North America.)

For much of their lives, Arctic charr tend to school. However, as the spawning period approaches, they become territorial, with larger specimens, particularly mature males, defending optimal spawning gravels in which they coax female fish to cut redds and deposit their eggs in much the same way as Atlantic salmon and brown trout.

4.2 The life-cycle of the Arctic charr

As spawning approaches, the coloration of the Arctic charr intensifies dramatically, the silver colour of deep-water and sea-going fish deepening to orange, through a range of reddish hues to bright red and finally to deep vermilion. Furthermore, the leading edges of the lower fins, and a fold of skin under the upper jaw, turn white. Also, like Atlantic salmon and brown trout, male fish tend to develop a kype on the lower jaw with which to defend their spawning territories.

The timing of spawning varies with climate, often related to a preferred water temperature of 4 °C. This can be as early as September or October in the coolest

Fig 4.2 A. F. Lydon's painting of a torgoch and Alpine charr from the Rev. W. Houghton's *British Fresh-water Fishes* (1879).

northern regions, but is generally considerably later further south. Mature charr seek out suitable areas of gravel or broken rock in the beds of flowing water, either in stretches of river or stream, or else in well-aerated zones of deep water on a lake bed that are not prone to freezing in winter ice. Whilst some lake populations may spawn on the lake bed, many, including the stock in Windermere in the English Lake District, run tributary streams in which they lay their eggs before returning back into the lake's depths.

Sea-going charr generally return to their natal river systems, equipped through evolution by the same homing instincts as salmon and sea trout. They also share a tendency for some individuals to run different rivers as an insurance against local circumstances eliminating charr populations from affected catchments. The spawning fish, both landlocked and sea going, avoid silty areas, which might suffocate their eggs. In this substrate, the female scoops out a shallow redd roughly the size and length of her own body using her fins, and into this she releases between 3,000 and 7,000 eggs (broadly in proportion with the 1,750 eggs per kilogram body weight generality for the salmonids) whilst, simultaneously, the cock fish releases his milt. The female than covers the eggs by fanning gravel back over them, often digging another redd in the process and repeating this performance a number of times until she is spent.

The eggs incubate throughout the winter at temperatures often close to 0 °C, with temperatures above 8 °C proving fatal. The eggs take between 400 and 450 degree days to hatch, shorter than Atlantic salmon and perhaps reflecting an adaptation to cooler waters, with hatching of any batch of eggs occurring over a period of a few days. The alevins emerge at a body length of 18–20 millimetres generally around the first week of April, though this varies between locations and from year to year with light and water temperature. Thereafter, the alevins remain in the redds feeding on their large yolk sacs, akin to all other salmonids, for a further 280–300 degree days (typically 5–6 weeks) before emerging as free-swimming fry. Fry swim-up is generally synchronised with the springtime emergence of the plankton on which they feed.

Consistent with their adaptation to cool water, growth is slow, despite flexible habits and a diet of small fish. After one year, Arctic charr are generally less than five centimetres (two inches) long. Growth rate thereafter varies dramatically with

Fig 4.3 View of Snowdon from Llyn Cwellyn, home of the torgoch.

habitat, but individual specimens can reach an age of 30 years and grow as heavy as 15 kilograms (33.1 pounds). Where they thrive, Arctic charr shoals can be dense. However, the charr is one of the rarest fish species in Britain, found only in deep, cold, glacial lakes, mostly in Scotland. It is also found in deep mountain loughs in Ireland. In the few English and Welsh lakes in which charr are found, they are often endangered by nutrient enrichment (eutrophication), which promotes blooms of algae that subsequently decompose, robbing oxygen from the cooler, deeper layers of water in which charr shoal and feed.

In other parts of its range, such as Scandinavia and into Russia, Arctic charr are much more common and are fished for extensively. Owing to these exacting requirements, Arctic charr are at risk from many human pressures, including not only the eutrophication effects noted above but also acidification and the silting of spawning gravels.

4.3 The value of Arctic charr stocks

The charr fishery on Windermere has a history going back to at least the mid-nineteenth century. The origins of the fishery are obscured by time, with clear records going back only to the late 1840s, but rumour suggests that medieval monks also fished the lake for charr.

In the latter half of the nineteenth century, an increasing demand for charr led to serious over-fishing and a reduction in catches and average size. The mesh sizes of the nets were reduced until they were catching 'minnow-sized' charr, in addition to salmon smolts, which displeased the salmon fishermen on the River Leven. Some of the proprietors of the fisheries and then the local Board of Conservators (after it had been set up by the Fisheries Act of 1865) sought to control the fisheries. Eventually the Board bought up all the netting rights and stopped all netting in 1921, though a characteristic semi-commercial plumb-line charr fishery has continued to this day.

This Windermere line-based charr fishing is an art and a tradition all of its own. However, like so many wild salmonid fisheries today, it is one that has declined drastically in the past generation, along with the fortunes of the charr populations themselves. Windermere charr boats are distinguished by two ash or, latterly, bamboo poles arching away from the stern, serving as outriggers for the tackle fished deep. Amidst the rapid pace of modern world activities around the lake shore, the charr fishermen paddle their boats sedately, working the lake's surface along traditional routes passed down over generations.

Deep below, spinning tackle works away. This comprises a set of metal spinners attached at varying depths, connected with specialised swivels to a main line weighted down by a lead weight of one and a half to two pounds. Often, indicator bells are attached to the tips of the charr poles to signal a taking fish. This specialist approach seems to have evolved uniquely to the English lakes, centred on Windermere but replicated at different scales in other Cumbrian lakes

Fig 4.4 Windermere, home of Arctic charr and a charr fishery.

including Coniston, Crummock, Ennerdale, Haweswater, Buttermere, Wastwater and Ullswater.

Rather later in the history of the Windermere charr fishery, net fisheries were established with up to ten licensed boats working the lake in its heyday. However, the Windermere net fishery was never particularly intensive nor long lasting, although each boat could take hundreds of fish on a good day. At its peak in the late nineteenth century through to the 1920s, fed principally by the pole fishers and augmented by netsmen, thousands of pounds of charr were harvested from Windermere every year, many of which, without freezing facilities, were potted (cooked and stored in solidified butter) and shipped to London.

The principal reason for the decline of the Windermere fishery, and indeed that of many English lakes, is related to the eutrophication-driven suppression of oxygen concentrations in the deeper lake water in which Arctic charr spend virtually all of their lives. Add to this the introduction and explosion of populations of coarse fish, particularly roach, many of which may have been introduced inadvertently as pike anglers released live bait brought to the lake, and the competition for lake habitat and predation on charr eggs and fry has become increasingly intense.

Farming of Arctic charr was slow to take off, despite early attempts around 1900 to fertilise and hatch their eggs. The slow growth rate is not ideal for farmed animals, but the adaptations of charr suit them to colder water temperatures. This fostered increased interest from the late 1980s, particularly in Iceland, where high-quality fresh or brackish water flowing through ponds is sometimes augmented with geothermally-warmed water to speed up the life-cycle and enhance growth. The number of Icelandic Arctic charr farms increased up to forty by the early 1990s, though profitability was marginal, resulting in several farms going out of business. However, interest subsequently revived, prompted by a government breeding programme initiated in 1992, and twenty Icelandic farms were producing Arctic charr in 2007. Today, Iceland is the world's largest producer of Arctic charr, accounting for more than half of global (in practice northern hemisphere) production.

Elsewhere across its native range the Arctic charr is far more important as a commercial species. For example, it has always been an important part of the food catch of the Inuit peoples, who catch charr with gill nets, traps and spears. From about the 1860s, charr have been exploited commercially on a very small scale in Labrador, Canada, but this burgeoned into a significant industry following government sponsorship of the processing market from the 1940s. The adaptation of charr to cold conditions means that they reach commercial size only after seven or eight years in the southerly part of their range, and up to fourteen years in the north. These slow growth rates mean that there is a persistent risk of over-fishing, with close monitoring required to ensure that fisheries remain viable and sustainable.

Fig 4.5 Rigged and ready for spinning for torgoch in deep water.

During much of the history of angling, recreational Arctic charr fisheries have been considered of secondary importance compared to trout. In the British Isles, at least, they are far more localised and require the more specialist approach discussed above to lure them from lake depths. However, to their aficionados, they are a valued and worthy prey. Across their wider geographical range, particularly the cooler climate of Canada, Scandinavia and Russia, they are more of a primary angling target much prized for both their sport and their flavour.

4.4 The role of Arctic charr in aquatic ecosystems

Their tolerance of cold conditions means that Arctic charr are one of the main fish predators in the cool regions of the northern hemisphere, in which they thrive in the absence of significant competition. This means that they play an important role in food webs, preying mainly on smaller fishes and, in turn, depositing nutrients into the ecosystem as they migrate into tributaries to spawn, and as they die or are eaten by large piscivorous birds, fishes and mammals. In practice, however, the Arctic charr has few enemies. Gulls and grebes may prey to a certain extent on small fish, whilst sea-going charr may fall prey to seals and whales.

Owing to the depleted fish fauna of the cool waters preferred by these fish, cannibalism accounts for much greater losses, and competition for food in nutrient-poor waters may play the most important role in regulating Arctic charr populations.

The various isolated and increasingly threatened populations of Arctic charr across the British Isles are recognised as priorities for conservation by inclusion in the UK Biodiversity Action Plan, giving rise to localised conservation measures to protect the species.

CHAPTER 5

The grayling

O ne key way in which the grayling differs from Atlantic salmon, brown or sea trout and Arctic charr is that it has no migratory life phase. Spawning also takes place in the spring rather than the winter. European grayling, *Thymallus thymallus* (Linnaeus, 1758), occur naturally from England and France eastwards throughout northern Europe and as far east as the Ural Mountains of north west Russia. Owing to the lack of any sea-going life stages, grayling have never penetrated into Irish waters.

5.1 Key features of the grayling

Across this range, grayling prefer running, well-oxygenated waters, and so are typical of clear, upper reaches of rivers with sand, gravel or rocky beds. In colder climates, particularly in parts of Alaska and Scandinavia, they occur in cool, well-oxygenated lakes, and there are some brackish-water populations around the Baltic Sea. Their water quality requirements are exacting, so grayling are extremely sensitive to pollution.

From their graceful appearance, streamlined and with silvered, iridescent flanks dappled with irregular dark spots interspersed with hints of purple, green and copper, and with a prominent sail-like dorsal fin, grayling sometimes delight in the title of 'Lady of the Stream'. This grand dorsal fin, mottled with black and red bars and raised at times of stress or to 'sail' in strong currents, also indicates the gender of the fish. Male grayling have the largest and deepest-hued fins, which they use to display to each other and to females at spawning time. The mouths of grayling are toothless and well adapted for feeding predominantly on the bottom, with the top lip extending beyond the lower. This perhaps explains why grayling feeding from the surface tend to roll and splash as they intercept floating food items.

Another curious fact about the grayling is the smell of a fresh fish, often likened to fresh thyme, from which the species derives its Latin name *Thymallus thymallus*. As Izaak Walton's 'Piscator' notes of the 'Umber and Grayling' in *The Compleat*

Angler, 'And some think that he feeds on water thyme, and smells of it at his first taking out of the water . . .' We know that this suggested diet is implausible, as grayling feed all year round on insects and other invertebrates and not just in the growing season of streamside herbs! However, the fact remains that grayling do possess a characteristic odour.

Grayling are gregarious, often found in small- to medium-sized shoals. Their diet comprises mainly aquatic insects and their nymphs and larvae, small worms and crustaceans, and they continue to feed and remain active in cool conditions, making them an attractive target for winter game and coarse fishers alike.

5.2 The life-cycle of the grayling

Grayling spawn typically between March and April depending upon local climate, and when the water temperature exceeds 4 °C. They are communal spawners, though male fish defend territories within a spawning group of fish. In the heartland of their range, grayling tend to spawn on suitable gravels close to the habitats in which they spend most of their lives. However, in colder regions, spawning migrations up river may occur following the spring thaw. Coastal populations in the brackish Baltic Sea around Sweden may either migrate into rivers to spawn, or else will lay their eggs in gravels along the shore of the northern Gulf of Bothnia.

As is typical with salmonids, male grayling attract gravid females into their spawning territories. However, for grayling, observations suggest that this behaviour may sometimes be diurnal, male fish leaving the spawning gravels at night but returning in the morning to reclaim a territory. At spawning time, the body colour of the male fish darkens and intensifies, the spectacular iridescent blue-red dorsal and pelvic fins being held erect to display to females and entice them to dig redds and lay their eggs. Male fish use their prominent, brightly coloured dorsal fins to attract mates and may hold a willing female close by curling it towards her prior to egg laying. Most spawning occurs as water temperatures peak in the early afternoon. The redd dug by the hen fish in well-oxygenated river gravels is shallow, but a forty-five centimetre (eighteen

Fig 5.1 A. F. Lydon's painting of a grayling from the Rev. W. Houghton's *British Fresh-water Fishes* (1879).

inch) fish may deposit as many as 10,000 sticky yellow eggs into it, making them more fecund than many other salmonids. The eggs hatch after 180–200 degree days, which equates typically to three or four weeks in the spring period in temperate climates. Alevins remain in the redd until the yolk sac is consumed then develop into fry which 'swim up' to live thereafter as free-swimming fish.

Grayling mature at an age of three or four years, which is quick for a fish of cooler climates. Their high fecundity means that populations can recover quickly from any dip in numbers.

5.3 The value of grayling

There are no records of commercial grayling fisheries, nor of grayling aquaculture, though the fish is frequently taken for the table across its geographical range. Historically, grayling have also suffered something of a lower-order reputation in angling circles. Because grayling spawn at the same time as coarse fish, their angling season corresponds to that for coarse fishes in the British Isles, though they are still legitimately 'game fishes' of the salmon family. Perhaps for this reason, grayling have unfortunately been treated as a 'lesser' species throughout

Fig 5.2 A grayling viewed from underwater.

Fig 5.3 In addition to the fly, grayling respond well to coarse angling methods with trotted maggots particularly effective.

history in the same way as coarse fish, although they have enjoyed a resurgence of interest in recent years in the UK, due largely to the efforts of the Grayling Society, and are now rightly regarded as a prized quarry in their own right.

Whilst grayling respond well to coarse angling methods, particularly trotted maggots, small red worms or sweetcorn, they will also eagerly take a fly fished dry or wet, especially as a nymph. They have the distinct angling virtue of feeding reliably in exceptionally cold conditions, during which time many other species of freshwater fish are dormant, and so can be fished for outside of the game fishing season. Indeed, Boxing and New Year's Day forays for grayling are something of a tradition in some parts of England.

For all the snobbery that may have subdued the status of this charismatic salmonid, few doubt its culinary virtues. As Izaak Walton's 'Piscator' notes in *The Compleat Angler*, '. . . he is accounted the choicest of all fish. And in Italy, he is, in the month of May, so highly valued, that he is sold there at a much higher rate than any other fish'. The culinary and sporting value of the grayling has been long and widely appreciated wherever this graceful fish swims.

5.4 The role of grayling in aquatic ecosystems

Grayling have some of the most exacting water quality requirements of any freshwater fish species, thereby serving a valuable role as an indicator of pollution (much as the canary was historically used to indicate foul air in mines).

Also, grayling often occur in the same river reaches as brown trout and the parr of Atlantic salmon and sea trout, and have historically suffered persecution for their perceived competition for food with these more prized fishes, as well as their keenness to intercept an angler's fly intended for their fellow salmonids.

In fact, the underslung mouth of grayling suits them to a subtly different diet to that of trout and salmon, enabling them to probe between stones on the river bed. Furthermore, adult grayling tend to select deeper runs than the shallower glides and riffles favoured by younger trout and salmon, perhaps reflecting an evolutionary adaptation to avoid competition and maximise the potential for

Fig 5.4 A grayling in Wiltshire's River Wylye.

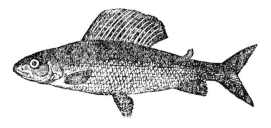

Fig 5.5 Woodcut of a grayling.

coexistence between the species. This rule of thumb may, however, break down where grayling have been introduced beyond their natural range by man, and no such coevolution with native trout populations has taken place, such as in some waters in Sweden.

Nevertheless, the resurrection of the grayling from perceived vermin to treasured angling quarry and indicator of high-quality water is as welcome as it is long overdue.

Furthermore, the conservation value of the grayling is recognised internationally by their inclusion together with Atlantic salmon in Appendix III (requiring controlled exploitation) of the Bern Convention. Grayling are also included under Annex V (management measures to control exploitation) of the EU Habitats Directive.

Base image courtesy of Dr Ian Winfield,
Centre for Ecology & Hydrology

Chapter 6

The whitefishes

In addition to the four game fishes that are the primary focus of this book – Atlantic salmon, brown or sea trout, Arctic charr and grayling – British members of the salmon family also include the whitefishes. This chapter provides a brief overview of these less-common species, though they will not feature any further in our consideration of the British game fishes.

6.1 Key features of the whitefishes

The whitefishes, also known as the coregonids, belong to the subfamily Coregoninae. Their inclusion in the salmon family is revealed by features including the fleshy adipose fin between the dorsal fin and the tail. However, they have small mouths and weak teeth, separating them from the salmon, trout and charr (subfamily Salmoninae), whereas teeth are lacking entirely in the grayling (subfamily Thymallinae).

Whitefishes are nevertheless important creatures of deep coldwater lakes, where they are also a major food species for various other game fish. Some whitefish species also inhabit rivers and even brackish waters across their broad circumpolar range, including in the Baltic Sea. Their general appearance is herring-like, with large silvery scales on the body and scaleless, relatively small heads. The British species have had a contested taxonomy, but are now considered to include the houting, the vendace and the European whitefish.

The houting (*Coregonus oxyrinchus*), which possessed small and feeble, practically toothless jaws, and inhabited the sea and estuaries, is now declared extinct and so will concern us no more.

The vendace (*Coregonus albula*) is superficially herring-like, growing to 48 centimetres (19 inches) and a weight of 1 kilogram (2.2 pounds). Vendace form pelagic schools in deeper lakes, feeding largely on planktonic crustaceans. With an elongated but laterally-compressed body, these fish have small mouths, sometimes with minute teeth on the jaws or tongue, or both. Their 34–52 long, well-developed gill rakers, and a lower jaw that fits in a narrow groove in the

Fig 6.1 A. F. Lydon's painting of a vendace from the Rev. W. Houghton's *British Fresh-water Fishes* (1879).

upper jaw and also protrudes slightly beyond it, are characteristic features that equip the vendace for life as a planktivore in deep, cool-water lakes. However, brackish and estuarine populations occur around the Baltic Sea.

The European whitefish (*Coregonus lavaretus*) is a cold-water species inhabiting deeper regions of large lakes. Also superficially herring-like, these whitefish grow to 46 centimetres (18 inches) and a weight of 0.9 kilograms (2 pounds) and have a conical head, a ventral mouth and 35–39 fairly long and well-developed gill rakers. The maxillary bone reaches to the front of the eye or only just past it. The dorsal, pectoral and pelvic fins are large and conspicuous when compared with vendace, to which this fish is closely related. The back is bluish or dark green, often very dark, and the dorsal, anal and tail fins are also dark. The tips of the pectoral and pelvic fins are dusky, sometimes for up to half their lengths.

European whitefish feed largely on invertebrates both on the bed and in the water column, with some populations known to rise up in the water to feed by night. With a laterally compressed, spindle-like body, these fish have small mouths sometimes with minute teeth on the jaws or tongue, or both. The European white-fish is also known as the schelly or skelly (in Haweswater, Ullswater and Red Tarn in the English Lake District), powan (in Loch Lomond), gwyniad (in Llyn Tegid [Lake Bala] in Wales) and pollan (in northern Ireland). However, the Welsh popu-lation in Lake Bala is considered by some authorities as the separate species *Coregonus pennantii*. Likewise, the Scottish population, the powan, is endemic only to Loch Lomond and Loch Eck and is considered by some authorities to be a

Fig 6.2 A. F. Lydon's painting of a pollan and a powan from the Rev. W. Houghton's *British Fresh-water Fishes* (1879).

separate species *Coregonus clupeoides*. Some also continue to regard the pollan of Northern Ireland to be a separate subspecies, *Coregonus autumnalis pollan*, or else a separate species *Coregonus pollan*. What is certain is that the European whitefish is of variable form across its range. Being somewhat more tolerant of warmer water than other whitefish species, the European whitefish is the most widely spread whitefish in western Europe and is the only one to be found in shallow lakes.

6.2 The life-cycle of the whitefishes

Relatively little is known about vendace biology, as these fish inhabit deep, well-oxygenated water for most of their lives. However, they come into shallower waters between 3 and 10 metres deep around lake shores to spawn in the autumn and winter (November to January) on sand or gravel substrates. However, there are spring-spawning populations of vendace in several North European lakes, some authorities considering these to be a separate species. Adapted to cool waters, growth rate of juvenile fish is slow. Vendace can live for up to 10 years.

European whitefish breed from October to January and, exceptionally, into March, spawning on marginal gravels on the shoreline of the large lakes that they inhabit when the water is around 6 °C. Also adapted to cold waters, growth rate is slow with juvenile fish attaining 6 centimetres (2^1/$_2$ inches) in their first year. European whitefish can also live up to 10 years old.

Fig 6.3 Llyn Tegid (Bala Lake), Wales, home to the gwyniad.

All whitefish require good quality, well-oxygenated water in deep lakes. Threats include pollution, particularly by organic matter and nutrient enrichment, which robs the deeper water layers they inhabit of oxygen. Sedimentation of spawning grounds has also been identified as a major conservation concern for the vendace, compounded by the introduction of coarse fish species, including roach and ruffe, which consume the vulnerable eggs and fry and are implicated in the extinction of vendace from Bassenthwaite Lake in the English Lake District. Similar threats have been recognised for the Welsh population of European whitefish (gwyniad) in Lake Bala, including the introduction of ruffe to the lake in the 1980s, which eat the vulnerable gwyniad eggs and fry. As a conservation measure, gwyniad eggs have been transferred to Llyn Arenig Fawr, a nearby cold lake. The Scottish population has also been successfully introduced as a conservation measure into two more sites: Loch Sloy and the Carron Valley Reservoir.

6.3 The value of the whitefishes

Across their broad holoarctic range, the whitefishes include valuable commercial species, though they are not exploited in the British Isles other than a small-scale commercial pollan fishery on Lough Neagh, Northern Ireland. Neither are they fished for with rod and line here, as all species and most local populations are subject to some degree of nature conservation protection.

However, these fishes are important as representative species of deep cold-water lakes, where they are considered glacial relicts stranded as ice cover retreated after the last ice age.

6.4 The role of whitefishes in aquatic ecosystems

Both the vendace and the European whitefish serve important roles in the nutrient-poor ecosystems they inhabit, consuming zooplankton and in turn being consumed by deep-water predators including Arctic charr, perhaps their main predator where the species coincide, as well as large brown trout (particularly the piscivorous ferox).

It is as species of conservation concern that the whitefishes receive most attention in the British Isles. The IUCN Red List of Threatened Species, a comprehensive inventory established in 1963 and regularly updated by the IUCN (the International Union for Conservation of Nature) of the global conservation status of species of plants and animals, recognises the gwyniad (of Wales) as Critically Endangered and the powan (Scotland) as Vulnerable. Both the vendace and the European Whitefish are also scheduled as priority conservation species under the EU Habitats Directive (Council Directive 92/43/EEC on the Conservation of

Natural Habitats and of Wild Fauna and Flora) as well as within the UK's Biodiversity Action Plan.

In addition, the Irish Government's Department of Environment, Heritage and Local Government has published All Ireland Species Action Plans (2009), addressing the four Irish populations in Lough Neagh, Lower Lough Erne, Lough Ree and Lough Derg.

Chapter 7

Rainbow trout: the familiar alien

Although this book is about the native British game fishes, the American interloper, the rainbow trout (*Onchyrhynchus mykiss*), first introduced to the UK in 1868, is now so widespread in our lakes and rivers that we have at least to give it a brief 'honourable mention', though it is clearly far from a native species.

The rainbow trout is generally hardier, faster-growing and more tolerant of moderate water quality than the native brown trout, which explains why it has been so extensively reared in aquaculture, both in the British Isles and worldwide to meet the needs of the food industry as well as to stock waters for recreational angling. Needless to say, stocked fish and escapees from trout farms have prospered in British waters, forming naturalised, self-sustaining populations in several of our rivers.

7.1 Key features of the rainbow trout

The rainbow trout is native to Pacific catchments of North America from Alaska to Mexico, inhabiting rivers and large clean lakes and feeding on aquatic and terrestrial invertebrates and small fishes. This trout is a member of the salmon family, though more similar in some regards to the charr than the Atlantic salmon.

The body form is elongated and streamlined, though increasingly laterally compressed in larger specimens. There is a characteristic wide pink to red stripe along the flank from head to caudal base, although this is lacking in the sea-running steelhead and also in the cultivated 'blue trout' form. Typically, spotting continues onto the tail, unlike the brown trout. The rainbow trout also lacks spines on the leading edge of the dorsal and anal fins. This fish grows to a maximum recorded weight of 25.4 kilograms (60 pounds), and can attain an age of 11 years.

7.2 The life-cycle of the rainbow trout

In all essential details, the life-cycle of the rainbow trout is similar to that of the native brown trout. This includes the sea-going form of the rainbow trout, the steelhead, which may now be establishing itself in the British Isles. Sea-going rainbows are certainly being caught in UK rivers, although whether these can be considered as genuine steelheads remains a matter of conjecture.

Rainbow trout are known to undertake short spawning migrations into suitable streams with clean gravel bottoms, the sea-going steelhead undertaking larger migrations from the sea into fresh water. Spawning behaviour is typical of the salmon family. However, most British rainbow trout still arise from stocking and farm escapes, with an increasing tendency towards stocking with sterile triploid fish, as described below, to limit environmental risks.

7.3 The value of rainbow trout stocks

It is as a food fish and recreational angling target that the rainbow trout is most appreciated, and these are indeed the reasons for its introduction to the British Isles as well as worldwide. Arguably, with its potential to disrupt locally adapted ecosystems and populations of native fishes, rainbow trout may have associated environmental costs that we do not currently count.

7.4 The role of rainbow trout in aquatic ecosystems

The rainbow trout is elegantly adapted to the Pacific coastal river ecosystems with which it coevolved. However, it is as a predatory species that the rainbow trout is of greatest concern where it has been introduced, predating on other species of potential interest as well as competing with native fishes. Adverse environmental impacts are reported from several countries. In the UK, trade in and keeping of rainbow trout are controlled under the Import of Live Fish (England and Wales) Act 1980.

Recognising some of the risks potentially associated with this and other introduced species, there is an increasing trend towards stocking triploid rainbow trout (sterile fish with three sets of chromosomes), which not only grow faster, benefiting aquaculture and angling interests, but which also depend on stocking to maintain populations.

CHAPTER 8

Realising the value of the British game fishes

To consider a wild fish separately from the waters that support its diverse needs throughout life and the seasons is to miss the whole essence of the animal. Fish, people, trees, microbes and organisms of all types are of the place of which they are part. Coevolved with the forces that have shaped the Earth and the diverse nature of the landscapes and the water bodies upon it, and the host of other species with which those ever-changing habitats are populated, creature and environment are inseparable. The worth of a fish, then, as indeed any other living thing, relates to how humanity may make use of it but also to its own inherent value and, just as pertinently, its role in the place in which it exists and to which it is inextricably linked.

Across their native range, for example, the four native British game fish species are emblematic of waters in healthy condition. They are not only all highly susceptible to pollution, but also to impoundments or other obstacles to their migratory needs, loss of suitable habitat, disturbance of other parts of the food web supporting their needs, altered flows and a range of other environmental pressures largely wrought by the hand of man. They can be, and in many instances have been, adversely affected by over-fishing, both commercial and recreational, as well as imbalances of predators or habitat degradation over the extended networks of environments necessary to sustain their life-cycles. They are vulnerable as well to the impacts of introduced and invasive species of plants and animals, including other fish species. Also to declining flows resulting from abstraction for a variety of human uses that, in addition to reducing habitat directly, also robs waters of the capacity to flush silt from spawning gravels and to be naturally re-oxygenated. We will turn to this concoction of pressures, and the prognosis for our freshwater game fishes if they are not addressed, in Part 2 of this book.

However, one of the key lessons we must take from this consideration of the intimate connectedness of fish and place is that the very presence of healthy populations of native game fishes in river and lake systems intuitively indicates to people, whether anglers or not, that all is well with their local aquatic world. Conversely, a decrease in their populations from rivers where they were historically present should be a 'wakeup call' that habitat, water quality or other ecosystem degradation has occurred. Something has happened and, like the miner's canary, the impact on local game fishes should alert us that all is not well. Any responsible remedial action taken therefore has a wider benefit than just local fish species. Owing to their requirement for adequate flows of clean water and a diverse ecosystem to sustain their life stages, management and conservation measures aimed at protecting these charismatic fishes are highly likely simultaneously to benefit most other forms of water-dependent life, and, therefore, also the many ways that people benefit from a healthy aquatic environment.

8.1 Benefiting from the services of nature

The direct economic benefits stemming from our freshwater game fishes, both in terms of commercial fisheries and angling revenue, are reasonably well understood and can be locally substantial. However, we are living in more enlightened times than those of our industrial past, which saw value only in the direct and generally detrimental exploitation of nature and natural resources, ranging from forests to fisheries, mineral, fossil carbon and metal deposits, water and so on. We are, in fact, progressively coming to recognise that intact and functional ecosystems support many different attributes of human wellbeing, many of which have been almost completely taken for granted, yet which, if lost, would have potentially serious negative consequences for those relying upon or enjoying them.

Mechanisms to classify these many benefits that ecosystems confer upon society have been developed progressively since the 1980s, and are now referred to as 'ecosystem services'. A classification scheme developed by the Millennium Ecosystem Assessment, a United Nations study initiated at the turn of the millennium to assess the status of global ecosystems and their consequences for human wellbeing, identifies four major categories of ecosystem services: 'provisioning', 'regulatory', 'cultural' and 'supporting' services.

Provisioning services relate to tangible goods provided by ecosystems. The commercial and informal harvesting of Atlantic salmon, brown or sea trout, Arctic charr and grayling clearly falls into this category, providing not only food but also associated livelihoods, which may be of considerable economic importance for local communities. However, the vitality of ecosystems capable of supporting thriving game fish populations also provide multiple additional 'provisioning' benefits to people, including yield of fresh water as well as the genetic, biochemical and ornamental resources of the diverse ecosystems upon which fishes depend and from which they are indissoluble. Lose the fish and we may also lose this whole treasure chest of provisions, many of which we have taken for granted throughout much of our history. The safeguarding of fish is not merely an act of altruism or self-interest advantageous to those who like to catch them for recreational or commercial purposes, but is also an investment in natural wealth for the long-term benefit of all of society.

Aquatic environments in a condition adequate to support game fish populations also regulate the wider environment in various significant ways. These range from the regulation of air quality and climate by habitats in the river corridor, including, for example, stands of riparian trees and reed beds that may trap aerial particulates harmful to human health, break down such dangerous pollutants as ozone and the oxides of sulphur and nitrogen, and store carbon. Riverside habitats may also play important roles in regulating water flows, for example with floodplains in their natural state absorbing and storing spates of water under heavy rainfall, retaining it such that flood peaks downstream are minimised, with an associated reduction in erosion of soils and damage to habitats and human infrastructure. This stored water is then released at a steadier rate, providing a more naturally buffered flow regime and so guarding against drought as well as flood. Indeed, modern approaches to flood-risk management are increasingly recognising the formerly substantially overlooked value of protected and restored riparian habitat for reducing flood damage, the ethos of 'working with natural processes' often now reversing historic drainage and development of floodplains, including the dismantling or setting back of engineered flood 'defences'.

These same river and riverside habitats may also be as important in regulating natural hazards, for example the absorption of storm energy by riparian trees, reed beds and other tall vegetation, minimising damage to ecosystems and crops as well as built infrastructure in the hinterland. Fringing wetland and river corridor ecosystems, essential to the vitality of the river channels and the fish populations within them, also support populations of predators that effectively regulate pests, and undertake decontamination processes that not only regulate disease organisms but also purify water. These habitats also, for the same reason, support populations of natural pollinators, potentially maintaining or increasing the productivity of crops and wildlife. Rivers and lakes with habitat suitable for native game fish populations thus serve society through a variety of important regulatory services that, if lost, would potentially have detrimental consequences for people.

We also derive a wide range of cultural benefits from aquatic ecosystems of a quality appropriate for native game fish populations. For many communities, river systems and corridors provide cultural heritage and regional distinctiveness, whilst the economics of angling and associated ecotourism, globally, nationally and locally, are highly significant both in terms of the angling activities themselves and related river management, tackle sales, travel and accommodation.

Fig 8.1 Trout fishing on a classic English chalk stream.

The contribution of native game stocks, and the ecosystems that support them, to the aesthetic and spiritual value of broader waterscapes and special river or lakeside sites are rather harder to evaluate. However, they may be very important to local people, communities and economies. As just one example, the legend of the 'Salmon of Wisdom' (*bradán feasa*) in Celtic mythology tells of a salmon that ate nine hazel nuts falling into the Well of Wisdom from the nine hazel trees surrounding it and, thereby, gained all the knowledge in the world, after which this knowledge passed to the first person (Fionn in the Celtic myth) to accidentally eat its flesh. The tale in Welsh mythology of how the poet Taliesin received his wisdom follows a very similar pattern, probably betraying a common ancestry. Elsewhere, salmon are the reincarnation of various Gods and famous people. The Salmon of Llyn Llyw is said to be the oldest animal in another ancient British tale, imbued appropriately with the wisdom of age. In Norse mythology, Loki (himself either a god or a jötunn, or giant, depending upon the telling of the myth) jumped into a river to escape punishment from the other gods, transforming himself into a salmon, but was eventually caught by the god Thor who grasped him by the tail, explaining why salmon tails to this day are tapered.

The salmon species of North America share a similarly rich mythology, much of which is central to Native American culture on the Pacific coast. Wider values associated with game fishes include their roles in religious ceremonies, peaceful retreats and, of course, the host of rather more familiar family picnics, walking, photography, bird-watching and other recreational activities related both to the fishes and the high environmental quality of the places in which they thrive, all of which add significantly to our quality of life.

Thriving waters hosting populations of native game fishes certainly provide inspiration for art and folklore locally. They are a focal point for social relations in commercial and recreational fishing, bird-watching and rambling, nature and heritage conservation, for local residents and other communities. Across Africa, central Asia and many other parts of the world where rivers run across national borders, the role of rivers, their water and ecosystems are a focus for international agreements and, in some cases, have served as an instrument central to peace-making between hostile or warring nations. Perhaps in microcosm, but no less important, the vitality and connectivity of ecosystems necessary to support thriving fish populations has been a useful focus for cross-border co-operation and agreement between countries, counties, towns and cities, landholdings and other political boundaries within the British Isles, focusing attention upon the health of the catchment ecosystem as a whole above more parochial concerns. Whilst British game fishes are clearly not themselves political activists, they are often a totem around which such important social agreements and capital are forged.

The vitality of ecosystems themselves, and the internal processes that maintain their integrity and the production of other ecosystem services, are summed up by the category of supporting services. These include important processes such as soil formation, generation of oxygen and primary production, provision of habitat for wildlife, and the cycling of water and nutrients. As we have seen in terms of the extraordinary life-cycles of salmon and trout, and their role in the remarkable life-cycle of the pearl mussel as well as interactions with a wide range of parasites, predators and prey, these fish may have important roles to play in the balance and food webs of ecosystems and the cycling of nutrients and energy. They are indeed indivisible from the environments that support them, and so indicate to us that the many often unappreciated, yet vital, supporting services upon which both ecological and human wellbeing depend absolutely, are happening for the benefit of all within aquatic ecosystems.

Since these various services provided by ecosystems reflect real or potential benefits to humanity, it is possible to ascribe monetary values to at least some of them. For instance, recreational angling, tourism and food production already have markets associated with them. Payment for angling licenses and other forms of access, travel and overnight accommodation, tackle and support for local businesses as part of the angling experience can and have been monetised, with much of this expenditure recirculated in the rural economy. Commercial fishing, too, has long-established market values, albeit that the market does not generally value the full environmental cost of fishing activities, which may without restraint be instrumental in significant harm to the ecosystems that produce the fish and, eventually, maintain the viability

Fig 8.2 Wild headwaters to which salmon and brown trout may migrate to spawn, just one of the many attributes of a valued pristine landscape.

of the fisheries themselves (as elaborated in Part 2 of this book). Water abstracted from clean and protected rivers and lakes also has market value. Other services, particularly the diversity of cultural and supporting services, are less easy to quantify, though nevertheless yield real benefits to society. For example, healthy aquatic ecosystems are significant and distinguishing features of the British landscape, partly monetised via tourism-related industries.

In some cases, it is possible to identify surrogate markets for these services, such as was contained in the report *Revealing the Value of the Natural Environment in England* prepared for the UK Government in 2004. This report assessed the annual value of the tourist trade attracted by a high quality natural environment in the UK as in the order of £5 billion in 2003, supporting the equivalent of 92,000 full-time equivalent jobs. However, this is far from adequate to account for the less tangible but no less important spiritual, regulatory, sense-of-place and other values of such treasured landscapes. Money is by no means everything, and we need to be ever mindful that the full palate of human enjoyment of the natural world, including the many often unappreciated facets that keep the natural world whole, functional and capable of sustaining humanity into the future, can defy such crude, man-made valuation devices.

Whether readily quantifiable or not, all of nature's many services to humanity are essential. If we take time to consider the many benefits to society stemming from the diverse services provided by healthy and functioning ecosystems or,

Fig 8.3 A restored reach of Devon's River Lyd, valuable for wildlife, angling, tourism and aesthetics, high-quality water and many benefits besides.

conversely, begin to account for the societal costs were those important ecosystem services to degrade or be lost, the magnitude of the value of healthy environments begins to become apparent.

Sometimes, the water requirements necessary to support ecosystems, such as ensuring minimum flow requirements to support different life stages of salmon, trout or other organisms of conservation importance, are perceived as competing with the needs and demands of people for water. So too, the costs of other conservation measures to retain environmental quality and diversity suitable to sustain the vitality of populations of species such as our game fishes. However, it is a fallacy to view ecosystems, in this blinkered and remote way, as representing a net cost or constraint upon human economic advancement. Ecosystems and their constituent biodiversity, including charismatic fish fauna, really matter both for their inherent worth and more tangibly in the many ways that they support the various dimensions of human wellbeing, from sustaining our basic biophysical needs through to providing economic resources and contributing to many less tangible aspects of our quality of life, achievement of potential and future security.

So, from this new, post-industrial, ecosystem-informed view of the world, what have our native game fishes ever done for us? The answer to this question is emphatic. The reality, through consideration of the multiple ecosystem services provided by the fish themselves and the ecosystems of which they are part, is that they have significant and often quantifiable benefits to all people, ranging from reliable yields of fresh water, valued landscapes, purification of waterborne wastes and the natural fertilisation and formation of soils.

Thus salmon, trout, charr and grayling are important not merely for anglers and commercial fishermen, but to all of society whether or not they be fishers or even that they consciously appreciate natural waterscapes. Rather than constraining society's progress, as nature conservation considerations are often critiqued by those with an outmoded narrow economic development mindset, fish are instead fundamental indicators of the natural wealth upon which future human wellbeing and opportunity depends.

Every last one of us is sustained by the vitality of the ecosystems that support the needs of fish and people, now and into the future.

British game fishes under pressure

CHAPTER 9

The making and breaking of the modern world

Part 2 of this book is all about the many and substantial pressures that our salmonid fishes face to survive and reproduce. Each of the following chapters deals with specific dimensions of threat, including those relating to introductions of fishes beyond their native ranges, fishing beyond sustainable limits, pollution and habitat modification in their various forms, the impacts of fish farming, and wider impacts affecting these vulnerable fishes. However, we also have to understand the broader-scale pressures exerted upon populations of salmon, trout, charr and grayling emanating from a world that has changed rapidly throughout human history, and will continue to change, immeasurably and increasingly in the future. This chapter sets that global and historical context, so as to better understand the set of pressures exerted on fish stocks, and the scale of the challenge of tackling them for the benefit of fishes, aquatic and other ecosystems, and humanity's dependence upon them now and into the future.

9.1 Tapping the wellspring

We, in the relative comfort of the Western world, are blessed by the double-edged sword of two-and-half centuries of industrialisation. On the positive side, this has yielded us expectations of longevity and health, personal wealth and material quality of life at scales not merely unprecedented but unimaginable throughout the sweep of human history. Those of us in the developed world, at least by sizeable majority, currently enjoy adequate – perhaps even excessive in some cases – access to food and drink, along with excellent education, healthcare and recreational opportunities. Our lives are no longer shackled by the requirements of daily survival, but provide us with opportunities for personal fulfilment that lie beyond even the most ambitious aspirations of many in the developing world.

However, the foundations upon which our emancipation is built have been systematically undermined by some of the very resource use, technology and

Fig 9.1 Our thirst for water is draining our waterways and wildlife dry.

lifestyle choices that have yielded these great advantages. Around the world, we have been systematically depleting natural systems and resources of fundamental importance to our continued wellbeing, economic activities and potential to live fulfilled lives.

From underground aquifers to wetlands, rivers, lakes and other freshwater systems, forests, soil and marine fish stocks, we have for too long treated the natural world as a free and boundless resource to be exploited for our own short-term ends. We have been blind to the harm that we have been inflicting upon it, and the associated longer-term consequences for our own security and wellbeing. All of these pressures have been exacerbated by a burgeoning human population, with increasing urbanisation and per capita resource demands.

9.2 People, people everywhere

The human population of planet Earth stood at only half-a-billion people at the outset of the Industrial Revolution. In the past century, the global population exploded to two billion by the 1930s, exceeding six billion by the turn of the millennium and seven billion by 2011. Demographers expect a billion more people to be added to the world population by 2030, almost all of this increase in cities in Africa, Asia and Latin America, then projected to reach a plateau of nine or ten billion by 2050. This overall trend has no precedent, and is furthermore a largely unstoppable consequence of current pressures.

Yet numbers alone do not tell the whole story. The volume and vitality of many natural resources are declining, leading to resource poverty and increased competition of all kinds, from water to food as well as industrial resources. Disempowerment, loss of creativity or denial of opportunity wastes human potential, sowing the seeds of social unrest and overexploitation of resources. Poverty can therefore be measured not only in economic terms, but also through access to resources, the equity of which has numerous indicators. For example, a 2005 study by the International Energy Agency records that 1.6 billion people lacked access to electricity, representing a quarter of the global population.

Meanwhile, hydrological poverty (lack of access to adequate safe water for basic needs and economic activities) is a particularly acute local form of impoverishment that is difficult to escape, and is becoming increasingly prevalent in a water- and climate-stressed world.

9.3 Trends in water

Planet Earth looks blue from space, largely because water covers two-thirds of its surface and pervades the atmosphere in the form of vapour. Of this apparent abundance, only 2.5% of the 1.4 billion cubic kilometres of water on Earth is fresh, all but 3% of which is permanently buried or frozen (much of this locked up in polar icecaps). The Earth's hydrological cycle constantly replenishes this fresh water supply. Seven billion people and much of the world's animal and plant life depend upon this renewable resource of just 0.075% of the Earth's water. Since much of the planet's biologically-active fresh water is seasonally or geographically inaccessible, more than half of what is actually available is used by humanity. As an essential solvent for all living processes, a sink for the residues of polluting activities, a key natural cycle and enabler of resource cycles, the state of the water environment is a prime indicator of the sustainability of our home planet.

Yet, remarkably, water is a resource we continue substantially to take for granted, despite recognition that it is already, and will increasingly become, a significant limiting factor to human development worldwide. Concerns about water quantity and quality are set to increase. Since 1950, global water use has more than tripled. In some areas, especially cities, rapidly growing populations are making demands on water far in excess of available supplies. Even when there is sufficient water, distribution infrastructure can be woefully inadequate, leading to inequities in access geographically, in terms of who is supplied and who is not, and the costs of obtaining it.

Human demand for water to meet more than just our basic needs is, in many ways, a key indicator both of our excessive and spiralling demands upon the natural world, and our seeming inability to share it equitably. An estimated 26 countries, with a combined population of more than 300 million people, currently suffer from water scarcity, including 9 out of the 14 countries of the Middle East. By the year 2050, projections suggest that 66 countries, comprising two thirds of the world population, will face moderate to severe water scarcity. Today, about 1.1 billion people, approaching a sixth of the world's population, do not have access to adequate clean drinking water, while 2.9 billion people lack satisfactory sanitation facilities. Some 1.8 million preventable child deaths each year are attributable to contaminated water.

Water consumption is also skewed to the privileged: the average US resident uses 400 litres per day, Canadians use 340 litres and a UK resident uses 150 litres

Fig 9.2 Booming human populations and industrial activity threaten river systems.

per day of which more than 50 litres is used to flush toilets. In addition to these figures, UK demands for imported food, textiles and other goods, often produced in and representing a net 'export' of water from developing and water-stressed countries, accounts for around 30 times as much 'virtual water' required to grow the timber, cotton, rice and other crops supporting our lifestyles, equating to a total UK per capita daily water demand of around 4,650 litres. Biofuel production exacerbates this still further, as these water-intensive crops require 1,000–4,000 litres of water to produce a single litre of biofuel.

This is compared to the 10–20 litres per day consumed by 1.8 billion people in developing countries who have access to a water source within 1 kilometre but not piped to their house or yard, with per capita daily consumption in some particularly remote and needy regions averaging as little as 3 litres. This may be substantially less than is necessary to support healthy lifestyles; deaths and debilitation are rife here, combining to keep the people of the developing world in a downward spiral of poverty.

The world is also placing rising pressures on surface waters and aquifers, with the advent of powerful diesel and electrically driven irrigation pumps during the latter half of the twentieth century now extracting groundwater often sub-stantially faster than natural recharge rates. This often means that water reserves are being used up or polluted faster than they can be replenished. As many as 2 billion people, approximately one third of the global population, and around 40% of the world's agriculture today, are reliant upon groundwater for drinking and irrigation and 10% of people's water consumption worldwide is from depleting groundwater.

Under the North China Plain, which accounts for 25% of China's grain harvest, the water table is falling by roughly 1.5 metres per year; probably much faster in some places. The same thing is happening under much of India, particularly the country's breadbasket in the Punjab, which is a major threat to food security. In the United States, water tables are falling under the grain-growing states of the southern Great Plains, shrinking the irrigated area. About 480 million people are being fed with grain produced by over-pumping aquifers. Drop-by-drop, water and food security are rapidly becoming eroded, with the changing climate exacerbating pressures on both.

Some of the world's major rivers are being drained dry and now fail to reach the sea, including the Colorado, one of the major rivers of the south western United States. In China, the Yellow River, the northernmost of the country's two major rivers, no longer reaches the sea for part of each year and, exacerbated by massive pollution, this contributed to the extinction of one third of the river's fish species by 2007. In Central Asia, the Amu Darya sometimes fails to reach the Aral Sea because it has been drained dry by upstream irrigation demands. The Aral Sea itself has shrunk beneath the relentless sun in this semi-arid region, its water level dropping 20 metres and its volume declining by half between 1960 and 1980, stranding ships in the encroaching desert as it split into two separate basins. By 2009, 75% of the original volume had been lost and the sea, once the world's fourth largest body of inland water, had receded into three separate lakes. Furthermore, sediment contaminated by pesticides accumulated on the now exposed former lake bed, blown by the wind, contributes to serious health concerns in towns that were once coastal but are now as much as 50 kilometres from the water. Were these recent trends to continue, the Aral Sea will largely disappear within another decade or two, becoming a geographic memory existing only on old maps.

At a wider geographical scale, there is a threat to the balance of water fluxes between continents and oceans. The silver lining to this cloud is that this actual and impending crisis has prompted concerted action by nations within the catchment of the Aral Sea leading to improved collaboration and water resource management in the post-Soviet era, the lake even beginning to refill to a small extent. But much damage has been done, some of it irreparable, and furthermore the resource depletion saga of the Aral Sea is being repeated around the world in the name of a pattern of development that fails generally to account for its wider consequences for aquatic ecosystems and the many people dependent upon them. There is far more to the planet's intricate water cycle than merely a commercially exploitable source of water for domestic, industrial and agricultural uses.

Natural ecosystems, especially sparsely populated uplands, wetlands and forests, capture water and stabilise seasonal flows, while recharging groundwater and improving water quality. Many are also 'hotspots' for both biodiversity and natural productivity. Conserving wetland ecosystems is vital for maintaining the supply of renewable fresh water, yet half the world's wetlands were lost to development during the twentieth century. And, as we over-harvest the scarce resource of groundwater, we also pollute it. UNESCO estimated a global $5 billion market in remediating contaminated groundwater and land throughout Europe, reflecting greater hidden costs from issues 'out of sight and out of mind', with the market for groundwater remediation set to escalate with as much as one third of the global human population facing water scarcity in coming decades. Pollution from agriculture, industrial and municipal sewage, and salinisation from irrigation, have also reduced the availability of clean fresh water. The global market for desalination stood at about US $3.8 billion in 2005 with a projected increase to $10 billion by 2015; not a sustainable solution when one considers the substantial energy demand of current technologies, but an indication of the magnitude of problems needing to be addressed.

We live in a water-challenged world, and more so each year as 80 million additional people stake their claims on the Earth's water resources. The situation promises to become far more precarious, since most of the two to three billion people that will swell the world's population by 2050 will be born in countries already facing water scarcity. With 40% of the world food supply coming from irrigated land, water scarcity directly affects food security and this, in turn, is compounded by competition with biofuel growth and land for development. If we are facing a future of water scarcity, we are also facing one of food scarcity, and that is increasingly threatening for wildlife and the ecosystems that provide many currently overlooked human benefits.

And so urgent action is needed. Over the past decade and more, water issues have received higher prominence. These include, for example, the resolutions of the World Summit on Sustainable Development (WSSD, Johannesburg, 2002) and the World Water Summit (Kyoto, 2003), the former formalising many of the Millennium Development Goals (MDGs) agreed by the developed-world community to contribute to the developing world in the new millennium. Water services underpin many of the MDGs, including a key target to halve the number of people without access to adequate sanitation (2.4 billion people in 2000) by 2015. The year 2003 was designated by the United Nations as the International Year of Freshwater, the aim of which was to focus attention on protecting and respecting our water resources, as individuals, communities and countries.

Although the UK does not face the severe water problems experienced in many other countries, there is certainly no room for complacency. Pressures such as new development, climate change and excessive domestic consumption are substantial and increasing, particularly in the south east of the country, where the water environment is already most significantly overstretched. Excessive water abstraction is

Fig 9.3 Our industrialised pathway of development has created many obstacles to fish migration, compounding wider problems.

one of the main stressors on once pristine river systems, including many chalk-streams, of which England plays host to about 90% of the global resource.

9.4 Biodiversity struggling on

If water is a key indicator of the effects of human development, and the 'space' available for humanity and all species sharing this planet, then the biological diversity of the Earth is a direct 'read out' of society's impacts upon its world. The omens are not good.

With each update of its 'Red List of Threatened Species', the World Conservation Union (IUCN) charts an increase in all categories of Critically Endangered species. The 2009 IUCN assessment revealed that 1,227, or one in eight, of the world's

Fig 9.4 Neither fish nor people can thrive without supporting ecosystems, indicated in British salmonid rivers by species such as alderflies.

bird species is in danger of extinction, further threatened by long-term drought and sudden extreme weather in the pockets of habitat upon which many threatened species now depend. In Europe 200 freshwater fish species (over one in three of the total) were recorded in the 2007 IUCN assessment as Threatened, and the 2006 assessment revealed that 56% of the 252 endemic freshwater Mediterranean fish were also under threat, representing the highest proportion of any regional freshwater fish assessment thus far recorded. The 2009 IUCN assessment recorded that 2,030 of the global 6,260 amphibian species assessed (nearly one third of species) are globally threatened or extinct, emphasising, together with the fish data, the extreme vulnerability of freshwater habitats.

Reassessment in 2007 of the status of the great apes, our nearest living relatives, made particularly grim reading, with substantial recorded declines. This includes the transition of the western gorilla (*Gorilla gorilla*) from Endangered to Critically Endangered status, already gravely threatened by the 'bush meat' trade, together with progressive habitat loss. This has been further decimated by a decline of more than 60% over the last 20–25 years, with about one third of the total population of one subspecies found in protected areas killed by the Ebola virus over the preceding 15 years. Habitat loss due to illegal and legal logging and forest clearance for palm oil plantations put increasing pressure on orang-utan species in Borneo. Chimpanzee species in West and Central Africa are threatened by similar bush meat and habitat loss pressures, with 97% of bonobos (the species most closely related to humans) disappearing in less than a human generation.

A major, yet commonly underestimated, threat to organisms of all types is the introduction of alien species, which can alter local habitats and communities, driving native species to extinction. Although this phenomenon is as old as human adventure, including the impacts of introduced rats on island ecosystems or the deliberate introduction of European foxes to Australia, the pace of species transfers around the world seems to be quickening dramatically. Alien, invasive fish introductions have had devastating consequences across the world. One outcome of globalisation, with its expanding international travel and commerce, is that more and more species are being accidentally or intentionally brought into new areas, where natural controls are absent, and are thriving to the detriment of established ecosystems. Of bird and plant species on the IUCN Red List, 30% and 15%, respectively, were recorded as threatened by non-native species in the 2000 assessment, and these figures appear similar in the 2009 assessment.

No one knows how many plant and animal species there are on Earth today, though current estimates range from 6 million to 20 million species, with the best working estimates falling between 13 million and 14 million. (It is impossible to fully quantify the biological effects of the most recent explosion in human economic activities, since we know even less about the number of species, and their global and local distribution, that existed a half-century ago.) We can measure losses where we have something approaching a complete inventory of species, such as for birds in countries with denser, better-educated and more environmentally aware human populations. However, only a fraction of species have been identified, described and catalogued for the many groups of small organisms, which number species in the millions, such as the insects. We understand far less about their roles within ecosystems, their potential as a source of pharmaceuticals or other resources, and the consequences of their loss. For example, we know virtually nothing about microbial diversity and function, beyond the fact that it is precisely these 'bugs' that are the major players in the functioning of global chemical cycles.

Best estimates quantify the current species extinction rate, heavily influenced by global human activities, as at least 1,000 times higher than the background rate. Irreplaceable species and ecosystem services, the products of 3.85 billion years of evolutionary fine-tuning, are being expunged by the march of technological progress over infinitesimally small time scales. As its diversity diminishes, nature's capacity to support the growing needs of humanity shrinks. The implications of global biodiversity losses for human wellbeing are serious and were the subject of detailed evaluation under the Millennium Ecosystem Assessment, established in 2000 by the UN to determine the status of global ecosystems and their prognosis for human wellbeing. Nature's larder, pharmacy, supply cupboard and cleansing service is depleted, depriving future generations of new discoveries, potential for sustenance, economic opportunity, heritage and 'quality of life'.

9.5 Vanishing forests

Forest cover is essential for the capture of water from clouds and rainfall, storage and purification of this water, smoothing of overland flows, and support for wildlife. Yet the 30% of the planet's land covered by forest or woodland is fast disappearing under clear-felling, burning, flooding and degradation by conversion to commercial forestry and intensive agricultural practices at manifestly unsustainable rates, barely dented by increasing forest cover and biomass in (historically denuded) Europe and North America. Worldwide, deforestation of natural forests in the tropics continues at an annual rate of over 10 million hectares per year, an area larger than Greece, Nicaragua or Nepal and more than four times the size of Belgium, although lack of clear data, and the inability from aerial images to distinguish mature forest from biodiversity-poor plantations, means that overall data are equivocal. The *Global Forest Resources Assessment 2000* of the UN's Food and Agriculture Organization reports that only 22 out of 137 developing countries possess time-series inventories, 28 countries have no inventory, and 33 have only partial inventories. Nevertheless, over-harvesting of forests is known to be common in many regions, particularly south east Asia, west Africa, and the Brazilian Amazon. After losing 97% of the Atlantic rainforest, Brazil is now destroying its Amazon rainforest, a huge resource roughly the size of Europe that was largely intact until 1970. Yet, by 2000, 14% had been lost, with 17,000 square kilometres deforested in 1999 alone. The past decade has witnessed fires on an unprecedented scale in the tropical forests of Brazil and Indonesia.

It is not just the developing world that has suffered from forest loss. For example, until the latter half of the nineteenth century, Sweden's industry was overwhelmingly forest based. It is inherent in the culture of managed forestry that one plants and manages for the long term, with stewardship as part of the ethos, unlike many other industries founded substantially on quick payback. However, around the end of the nineteenth and start of the twentieth centuries, Sweden faced two huge blows. First, over-harvesting of the seemingly boundless resource of trees had reached a point where the forests were in a severely depleted state, with drastic economic consequences. Second, at around the same time, the nation was hit by a devastating famine that claimed many lives. So Sweden is aware, from relatively recent cultural memory, that adversity born of ecosystem depletion can have serious consequences for human wellbeing and the economy. These forces may contribute to Sweden's enviable reputation for environmental literacy and progressive leadership for sustainability, albeit that further globalisation and industrialisation threatens this cultural legacy.

Fig 9.5 Trees provide essential waterside habitat but also play many roles in regulating the wider environment.

It is not merely the long-term security of the economically valuable asset of wood that is at issue when forest ecosystems are removed. Deforestation can be costly in many other ways. These include perturbation of the hydrology of whole watersheds. For example, the unprecedented flooding in the Yangtze River basin during the summer of 1998 drove 120 million people from their homes and, although initially referred to as a 'natural disaster' (reported in 'China Daily', 1st September 1998, and reviewed in the Worldwatch Institute's 1998 book *Watersheds of the World*), it was later recognised that the removal of 85% of the original tree cover in the basin had left little vegetative cover to hold the heavy rainfall. Not only was direct flood damage consequent from forest removal, but the surge of storm water, no longer slowed and stored by mature forests, carried with it huge quantities of topsoil, depleting the primary resource of agriculture and threatening to shorten the life and efficiency of water storage, energy generation and navigation in the massive Three Gorges Dam downstream.

Further harm to human interests flows from declining biodiversity and ecosystems consequent from these primary impacts. The same principle applies in peninsular India, where the forested mountain chain of the Western Ghats, bordering the western edge of the subcontinent, intercepts moist south-west monsoon winds blowing in from the Arabian Sea. Rainfall, water storage and purification are significant functions of the mosaic of wet and dry forests of the Western Ghats, constituting biodiversity 'hotspots' that are also essential for maintaining the flows of the three great and holy river systems of Deccan India – the Godavari, Krishna and Cauvery – and feeding moisture into the otherwise arid lands of the Deccan peninsula. Loss of the distinctive and functionally important habitat and biodiversity of the Western Ghats, which is proceeding apace, could ultimately be catastrophic for the supply of water to much of the subcontinent. The same principles also apply to broad-scale forest felling in the Amazon, South and eastern Africa, Indonesia and many other parts of the world. Water storage, recycling and purification, moderation of flood pulses, production of food and resources vital to sustaining livelihoods and a host of other environmental benefits besides, are likely to be the net casualties of forest loss, along with characteristic and valued biodiversity, from which long-term human interests inevitably suffer.

The UN *Millennium Ecosystem Assessment* records that forest ecosystems are extremely important refuges for terrestrial biodiversity, forming central components of the Earth's biogeochemical systems as well as a source of ecosystem services essential for human wellbeing. About two-thirds of the global population, 4.6 billion people, depend for all or some of their supplies on forest systems. However, both the area and condition of the world's forests have declined throughout recent human history. In the last three centuries, global forest area has been reduced by approximately 40%, with three-quarters of this loss occurring during the last two centuries. Forests were found to have completely disappeared in

25 countries, whilst another 29 countries have lost more than 90% of their forest cover. Since forest, particularly tropical forest, provides habitat for half or more of the world's known terrestrial plant and animal species, this represents a serious erosion of global biodiversity, and of the many beneficial functions provided by forest ecosystems. These include their role in carbon cycling, provision of resources for frequently poor dependent communities, and the many cultural, spiritual and recreational needs they service for different societies.

Forest loss and degradation does not just happen by chance, nor is it necessarily due to the bad intent of local people. It is driven by a combination of economic, political and institutional factors that primarily include agricultural expansion, high levels of wood extraction, and the extension of roads and other infrastructure into forested areas. Where present, industrial forest plantation subsidies, including income support that does not work for the best long-term interests of local people, or the construction of road and rail access into areas of old-growth forest, which stimulate legal and illegal exploitation, have frequently contributed to a range of foreseeable or unforeseen impacts upon the biodiversity and other facets of global forests.

Principles of sustainable forestry management have been developed to support more long-term economic exploitation that respects forest integrity together with the rights and needs of forest dwellers and those dependent upon the many services provided by forests. They also underpin independently audited certification initiatives, such as the excellent Forest Stewardship Council scheme to assure consumers of products from sustainably managed forests. However, the practice of sustainable forest management is still far from pervasive, and the destructive, short-sighted trends remain to be slowed, halted and, ideally, reversed by forest restoration, to recover the many essential, not to mention economically and otherwise vital, ecosystem functions lost to cavalier short-term exploitation and development.

9.6 Shrinking Earth, barren oceans and empty food baskets

As a consequence of short-term exploitation eroding long-term interests, the loss of topsoil from wind and water erosion now exceeds the natural formation of new soil over large areas of the world, gradually draining the land of its fecundity. These trends are converging to form increasingly large dust storms, such as the huge plumes that now commonly shade out the sun in cities of north east China, but which can contribute to higher-altitude aerosols entering and dispersing into the atmosphere, with potential implications for energy exchange. Further threats to the stability of soils arise as unintended consequences of mass deforestation, as already addressed in the context of China's Yangtze River basin, as well as widespread conversion of permanent vegetation cover to tilled land, particularly on hill slopes. Such 'hidden' environmental and social costs are referred to as 'externalities', as they are still largely excluded from economic valuation and decision making. For example, a case study involving Madagascan rainforests re-vealed that the cumulative values to society accruing from forest conservation sig-nificantly exceeded those stemming from logging and agriculture at both local and global scales, yet logging continued as this was driven by greater incentives at a national scale that did not recognise the wider societal benefits of intact forests. Such perverse economic 'signals' continue to exacerbate tropical deforestation, wetland and soil loss, and the depletion of other ecosystems, biodiversity and natural resources worldwide, even where we can already foresee the long-term harm they may produce. The principle applies equally in catchment and coastal management, where most of the benefits of good practice accrue to a range of users of catchments who may not pay for the actions that deliver them. More sophisti-cated economic tools are essential if sustainable management is to be realised.

It is commonly agreed that two-thirds of oceanic fisheries are now being fished at or beyond their sustainable yield. As a direct consequence, many are collapsing or have already done so. The FAO (Food and Agriculture Organization of the UN) report, *The State of World Fisheries and Aquaculture 2000*, and the book, *The Business of Biodiversity*, review the doleful tale of the Newfoundland 'Grand Banks' cod fishery. This rich fishery had been supplying cod for at least five centuries yet, in 1992, collapsed abruptly. This cost an estimated 30–40,000 Canadians their jobs, robbing many fishing communities of their livelihoods and seeding a culture of poverty, depopulation and domestic breakdown. Two decades on, and despite a ban on fishing all cod stocks, the fin fishery is only just showing signs that it might tentatively recover.

The addition of 80 million people a year to world population at a time when water tables are falling suggests that food supplies may be a vulnerable link between environment and economy. An estimated 70% of the water consumed worldwide, including that diverted from rivers and pumped from underground, is used for irrigation. Of the remainder, 20% is used by industry and 10% for domestic purposes. In the increasingly intense competition for water among these three sectors, the economics of water do not favour agriculture. Neither do they favour the continued existence of viable sources of water or the ecosystems that depend upon them, including their most vulnerable iconic and valuable fishes. The fact that long-term human wellbeing is ultimately wholly dependent upon these ecosystems is only now beginning to enter political calculations.

It is not just how much a growing population consumes that should concern us, but also what those people eat. A US diet, rich in livestock products, requires four times as much grain per person as a rice-based diet in a country like India. Using four times as much grain means using four times as much water and land, greater demand for energy, fertilisers and other agrochemicals, and pressures upon terrestrial ecosystems with widespread consequences for wildlife, soil health and fluxes of water through landscapes. As over-exploitation of water, ecosystems and soils compromise the viability and fertility of the land, so too these impacts affect the climate, and in turn the resilience of the land. We are literally planting the seeds of a vicious cycle, unravelling the very web of life upon which humanity depends.

9.7 A changing climate

Atmospheric carbon, in the form of carbon dioxide (CO_2), plays an important role in making this planet habitable. The term 'greenhouse effect' was first coined by Swedish scientist Svante Arrhenius, yet the effect itself was first predicted in 1827 by French mathematician Joseph Fourier. The greenhouse effect is a phenomenon of the Earth's atmosphere by which a proportion of solar radiation, reaching the inner atmosphere and re-emitted from the Earth's surface, is prevented from escaping by various gases acting much like the glass in a greenhouse. The main naturally occurring atmospheric 'greenhouse gases' are CO_2, methane and water vapour. The result is a rise in the Earth's temperature above that which could be predicted from its distance from the sun alone. This warming effect enables the existence of water in its liquid form on this planet, itself supporting the genesis and proliferation of living things. Nature's cycles maintain greenhouse gases in balance, contributing to the stability of the atmosphere.

Fig 9.6 A changing climate is placing growing pressures on British and global freshwater systems.

Today, atmospheric concentrations of a range of greenhouse gases are rising sharply. Fossil fuel consumption and forest fires are the main causes of CO_2 build-up, together with methane, a by-product of agriculture (predominantly from rice, cattle and sheep) and oxides of nitrogen. Prior to the start of the Industrial Revolution in the eighteenth century, anthropogenic CO_2 emissions from the burning of fossil fuels relative to natural background levels were negligible, and the atmospheric CO_2 concentration was estimated at 280 parts per million (ppm). By 1950, global emissions of CO_2 from fossil fuel burning had reached 1.6 billion tons per year, a quantity that was already boosting the atmospheric CO_2 level. In 2000, CO_2 emissions totalled 6.3 billion tons, boosting mean atmospheric concentrations to 370 ppm, representing a rise of 32% from pre-industrial levels, with the rate of increase estimated at 0.5% per year and rising. The build-up of atmospheric CO_2 from 1960 to 2000 of 54 ppm far exceeded the 36 ppm rise from 1760 to 1960. Atmospheric CO_2 levels have risen each year since annual measurements began in 1959, making this one of the most predictable of all environmental trends. The mean global atmospheric CO_2 concentration in December 2009 was 384.8 ppm and rising.

The concentrations of other greenhouse gases also give cause for concern. Chlorofluorocarbon levels are rising by 5% a year, and nitrous oxide levels by 0.4% a year. The ongoing situation is complex, with an estimate that the global rate of increase fell below this growth rate in 2008, due largely to high oil prices in the first half of the year and the economic slowdown in the second half, although increasing biofuel production also helped displace a substantial volume of fossil-fuel petrol and diesel. Nevertheless, accounting for 50.3% of global emissions in 2008, the developing world, which is seeing the major growth of greenhouse gas emissions due to burgeoning populations, increasing per capita material demands and implementation of dirtier, often outmoded technologies, exceeded the combined emissions of the developed world and international travel for the first time. In this complex and uneven situation, with demand for fossil fuels likely to rise with global economic growth including the industrialisation of developing nations, strong and concerted international leadership and action will be required if we are to avert serious consequences from unconstrained economic activities.

The Intergovernmental Panel on Climate Change (IPCC) was established in 1988 by the World Meteorological Organization (WMO) and the United Nations Environment Programme (UNEP), two organisations of the United Nations, to assess the scientific, technical and socioeconomic information relevant for the understanding of the risk of human-induced climate change. Aside from collating much of the information used in this evaluation, the IPCC has also been an exemplar in international co-operation to address a common threat requiring concerted international understanding and action. It thus quantifies the magnitude of threat, but also points to potential solutions, for which bold political leadership will be required.

The resulting rise in atmospheric CO_2 levels is widely believed by scientists to account for the warming of the world. The list of warmest years on record is dominated by this millennium, each of the years from 2000 to 2010 featuring as one of the warmest since these data were collated by the US National Climate Data Center (NCDC) temperature record. Reconstructions by the NCDC of earlier temperatures from more fragmented records suggests that these were also the warmest years for several centuries or millennia. WMO data reveal that 2011 caps a decade that set a record of being the warmest ever, between 0.7 °C and 0.9 °C warmer than any previous decade. This is substantiated in the UK, for which the Met Office recorded that 2011 was the second warmest year on record after 2006. Whilst the geological record shows that climatic changes have taken place regularly, most notably during ice ages, modern climatic changes are occurring at an unprecedented rate, and appear to be linked to increasing levels of pollution in the atmosphere. The United Nations Environment Programme estimates that, by 2025, average world temperatures will have risen by 1.5 °C, with a consequent rise of 20 centimetres in sea level. There are also implications for the disruption of important global sea currents, and the energy transfer and food production functions that they perform.

Rising temperatures lead to more extreme climatic events, including record heat waves, the melting of ice, rising sea level, and increasing frequencies of more destructive storms and intense rainfall events leading to major flooding. Further consequences include respiratory problems for humans in inner cities, loss of crop productivity and increasing risks to investments of all kinds. For example, the July 1995 heat wave in Chicago, when temperatures reached 38 to 41 °C on five consecutive days, claimed more than 500 lives and helped shrink the 1995 US corn harvest by some 15%, or $3 billion.

There was a recorded rise of about 1 °C in the temperature of the world's oceans during the 1980s. Arctic ice was six to seven metres thick in 1976 and had reduced to four to five metres by 1987. On 19th August 2000, the *New York Times* reported that an icebreaker cruise ship had reached the North Pole, only to discover this famous frozen site was now open water. The Arctic sea ice has thinned from nearly two metres thick in 1960 to scarcely one metre in 2001. Some scientists predict that, within 50 years, the Arctic Ocean could be ice-free during the summer.

In Europe's Alps, the shrinkage of the glacial volume by more than half since 1850 is expected to continue, with these ancient glaciers largely disappearing over the next half century. Predictions about the loss of ice and snow on Tanzania's iconic Mount Kilimanjaro are being realised, with increasing implications for tourism. Meanwhile, retreating glaciers in the Himalayan range in central Asia is threatening to reduce the dry season flow of the River Ganges by as much as 50%, with serious consequences for the millions of people dependent upon it. As 'the world's water tower', glacier and snow loss across the wider Himalayas, exacerbated by poor upland land-use decisions, may have serious consequences for major rivers and their billions of human dependants across the wider Asian continent.

All of this extra water has to go somewhere, and contributes substantially to the phenomenon of sea-level rise. For the first time since civilisation began, sea level has begun to rise at a measurable rate, threatening habitation of low-lying islands and land (including many major cities). During the twentieth century, the sea level rose on average by 18 centimetres, more than half as much as it had risen during the preceding 2,000 years and on a still-accelerating trend. Current sea-level rise has occurred at a mean rate of 1.8 millimetres per year for the past century and, more recently, estimates suggest an increasing rate of between 2.8 and 3.1 millimetres rise per year between 1993 and 2003. If current estimates are correct, the sea level could rise by as much as one metre during the twenty-first century, with the coastline retreating, on average, by 1,500 metres and some smaller islands becoming uninhabitable. In central London, the Thames barrier, comprising a set of ten movable gates constructed across the river in 1980 to prevent the upriver surge of tidal flows, has been closed with increasing frequency since its inauguration. Although compounded by the downward tilt of the land mass of south east England, the fact the barrier was shut three times a year on average in the first ten years of operation, but that this frequency has now doubled, provides some indication of the potential consequences of rising sea levels, exacerbated by increasing atmospheric energy contributing to storm surges.

A further consequence of higher temperatures is more energy driving storm systems. From 1920 until 1970, there were an estimated 40 major storms per year but, between 1985 and 2000, the northern hemisphere experienced close to 80 storms a year, representing a doubling in less than a generation. Publishing in the journal *Nature* in August 2009, a team of scientists from Penn State University determined that the annual number of Atlantic hurricanes is higher now than at any time in the last 1,000 years. With rising frequency has also come increasing force and consequent economic damage, such as the three powerful and damaging winter storms in France in December 1999, or 1998's Hurricane Georges in Central America. Hurricane Katrina, making landfall primarily in the state of Louisiana and resulting in the inundation of the city of New Orleans in August 2005, was one of the deadliest natural disasters in US history, leaving 1,836 people dead and a further 705 missing, with damage felt more widely, including an estimated $1–2 billion dollars' worth in the neighbouring low-lying state of Florida.

As a consequence of higher-energy storms and other phenomena relating to climate change, natural disasters are on the increase. Munich Re, one of the world's largest reinsurance companies, reported that three times as many great natural catastrophes occurred during the 1990s as during the 1960s, while economic losses increased eightfold and insured losses multiplied fifteenfold. Munich Re's natural catastrophe figures for 2007 demonstrate higher losses, despite the absence of mega-catastrophes, with cumulative events resulting in overall economic losses of US$ 75 billion, rising in line with the increasing trend in natural catastrophes. The CGMU Insurance Group (Britain's largest) reported in 'Environment News Service' (24th September 2000) that property damage worldwide was rising at roughly 10% a year. We are already seeing economic fallout from climate change, but it is projected to get much worse without serious intervention and, indeed, as a consequence of environmental pressures already in place. At this rate of growth, by 2065, the amount of damage would exceed the projected gross world product; well before then, the world would face bankruptcy. As Lester Brown, President of the Earth Policy Institute, suggests, 'Nature was levying a tax of its own on fossil fuel burning'. The UK Government's 2006 *Stern Review*, discussed in more detail below, highlighted the economic benefits of early action to avert the worst consequences of climate change, which would otherwise dwarf the costs of climate-derived damage.

An apparently small temperature rise in sea surface water of less than 1 °C can lead to the effect of 'coral bleaching', the mass death of corals, which has already devastated huge areas of reefs in the Caribbean and the Indian and Pacific Oceans. If the reefs continue to die, oceanic ecosystems will be altered, directly affecting the fisheries that depend on the coral reefs as nursery grounds. And as fisheries fail, in both tropical and temperate waters, we need to travel further to exploit remaining grounds, increasing the potential for conflict, driving the land ever harder to produce farmed crops to make up the shortfall in protein and other essential food groups, and all the while expending more energy, which feeds back in turn to climatic instability.

In addition to these diverse and worrying physical, meteorological and thermal effects, climate change is also exerting a range of other indirect pressures on the Earth's ecosystems. One of the most significant of these is ocean acidification, which is sometimes described as the 'evil twin' of climate change. Acidification of the oceans is brought about by elevated concentrations of atmospheric carbon dioxide, which increase the levels of dissolved gases in the oceans, thus decreasing their pH. Between 1751, which more or less pre-dates the onset of industrialisation, and 1994, the mean global surface ocean pH is estimated to have decreased from approximately 8.179 to 8.104. This apparently small net change in pH of -0.075 is equivalent to the degree of change throughout the last 300 million years. It is also responsible for a wide range of observed changed mineral flows and biological responses including, for example, making it increasingly difficult for animals to form shells, with various wider consequences for ecosystems. If pH falls by a further 0.3–0.4 units by the end of the century, which is likely on the basis of current unmitigated greenhouse gas emission trends, the consequences would be disastrous. Other diverse potential impacts of a changing climate include massively enhanced erosion associated with increasingly severe storms, droughts and other extremes of weather. This would result in the loss of bio-diversity and oxygen concentrations in aquatic ecosystems due to the effects of excessive nutrients carried into them by sediment particles, all as a result of more energy in the atmosphere.

At the 1992 UN 'Earth Summit' in Rio de Janeiro, it was agreed that, by 2000, countries would stabilise CO_2 emissions at 1990 levels. Progress was made through the Kyoto Protocol, an international agreement linked to the United Nations Convention on Climate Change (UNCCC). This set binding targets for 37 industrialised countries and the European community for reducing green-house gas (GHG) emissions, recognising that developed countries are principally responsible for the current high levels of GHG emissions in the atmosphere as a result of more than 150 years of industrial activity. These targets cumulatively amount to an average 5% reduction in total emissions against 1990 levels over the five-year period 2008–2012. Since the Kyoto Protocol, there has been emerging awareness and consensus, driven significantly by the work of the IPCC, about the magnitude of the threat posed by climate change and the need for substantial reductions in emissions if we are to avert its most extreme consequences. There is already consensus that global emissions would probably need to be cut by 60%, with the UK Government announcing a commitment in March 2007 that it would reach this target by 2050; this in response to a recommendation from its Royal Commission on Environmental Pollution in 2000. This UK commitment has since been upgraded to 80% in the light of subsequent scientific advice, with some authorities calling for a 90% cut in developed countries to allow the developing world to 'catch up' in economic and technological terms.

A follow-up showpiece UNCCC conference in Copenhagen in December 2009 fell disappointingly short of achieving serious commitments to limiting climate change gas emissions, and so placing collective wellbeing ahead of national self-interest. Contracting parties at Copenhagen failed to make any binding agreements, in the face of posturing from some industrialised nations and mistrust between the developed and the developing world. There were, however, pledges from the already-developed world, comprising countries that have historically benefited most from high emissions, to provide climate aid to developing countries. The formal outcome of the Copenhagen meeting, a weakened Copenhagen Accord drawn up on the concluding night, emphasised the need for strong political will to urgently combat climate change in accordance with the principle of common but differentiated responsibilities, and particularly recognising the urgency of holding global temperature rise to no more than 2 °C. So, though underwhelming, the outcomes from the Copenhagen conference were not all negative, albeit not quite transparently matched by country-level commitments proportionate to the threat to collective human wellbeing. A follow-up UNCCC summit in Durban in

December 2011 added little to this slow process of achieving concerted action, albeit that parties continued to negotiate into the early hours of the morning following the formal final day of the conference, failing to establish legally-binding targets but agreeing substantial financial support to help developing nations address climate change strategies. We can only hope that 'grassroots' unilateral action will continue along the lines of the many commitments worldwide entered into electively by cities, states, regions and nations, and that momentum will continue amongst businesses that realise the advantages of reducing their dependence on high levels of emissions. An inspiring example of state-level commitment, in the face of a federal failure to ratify the Kyoto Protocol and blocking tactics to achieve legally-binding targets, is the state of California which, formalised under the landmark California Global Warming Solutions Act of 2006, made commitments to address its contribution of roughly 1.4% of the world's and 6.2% of the total US greenhouse gas emissions.

9.8 The human cost

The conflict between the economy and Earth's natural systems can be seen in daily news reports of collapsing fisheries, shrinking forests, eroding soils, deteriorating rangelands, expanding deserts, rising CO_2 levels in the atmosphere, falling water tables, rising temperatures, more destructive storms, melting glaciers, rising sea level, dying coral reefs and disappearing species. This all raises questions of intergenerational responsibility that humanity has never before faced, at least not consciously. These trends mark an increasingly stressed relationship between the economy and the Earth's ecosystem, and they exact a growing economic and human toll.

The UN's *Millennium Ecosystem Assessment*, elements of which have been used to quantify issues discussed in this chapter, was instigated to determine trends in global ecosystems and their prognosis for continuing human wellbeing. The *Assessment* makes explicit that expectations for basic life support, economic

Fig 9.7 Dead fish and degraded environments have very real human consequences.

opportunity and the potential for people to live fulfilled lives depend entirely upon the services provided by the Earth's various ecosystems, suffering or prospering with their integrity and supportive capacities.

At some point these trends will, without radical overhaul of our economic and resource-use habits, overwhelm the capacity for further human progress and economic prosperity worldwide. The window of opportunity is narrow for us to change our dangerous trajectory of development. To remind us that this is more than mere rhetoric, we can look back to many earlier civilisations, such as Easter Island, the Aztecs and Incas, and a range of other civilisations across the world (as eloquently reviewed in Jared Diamond's 2004 book, *Collapse*) that outstripped the capacity of the environments that supported them, only to subsequently implode due to resource starvation or the vulnerabilities stemming from it. The difference is that never before has this threat occurred at the present scale, affecting the total global ecosystem, and within the understanding of contemporary science. The evidence is that we are a long way overdrawn upon nature's capacity to underwrite our survival and prosperity.

9.9 A common destiny

Coincident with growing societal awareness of the systematic degradation of the global ecosystem is a belated realisation that we are, in fact, all part of the same living global ecosystem, and that the protection of living resources is an urgent priority if we are to live secure and fulfilled lives. Without nature's capacities to purify air and water, detain floods and retain water resources, fertilise soils and produce food, support diverse ecosystems, and provide economic and aesthetic potential as well as places to live which may be of spiritual or cultural importance, humanity has an impoverished future.

All facets of global ecosystems, from charismatic birds and mammals to micro-organisms and fish communities, are essential for the future of the whole, internally-interdependent biosphere of this planet.

We share one world, at once fragile yet also endlessly renewable provided its natural limits are respected. We can choose collectively to allow it to provide for us indefinitely – or in other words sustainably – or else suffer together as we continue to overrun its supportive capacities and integrity. Ensuring access to basic resources and improving the health and livelihoods of the world's poorest people cannot be tackled separately from maintaining the integrity of natural ecosystems. In the face of high and rising population, allied with resource-hungry habits, this quest for sustainability is ever more pressing and ever more difficult, particularly with a present population which, by and large it seems, does not want to surrender short-term, fragile luxuries to face up to reality. This, in turn, takes pressure off governments to make difficult decisions, perfectly illustrated by recent UNCCC negotiations where, despite NGOs and protestors trying to make their mark, there was a marked dearth of political motivation to make responsible decisions to combat global climate change. It is hard not to make a comparison with Nero fiddling (he actually played the lyre as the violin had not yet been invented) whilst, around his ears, Rome burned as the empire started to crumble.

Sensible long-term policies seem elusive. For example, in 1994, the Chinese Government decided that the country would develop an automobile-centred transportation system. However, if Beijing's goal were to materialise, and the Chinese were to have one or two cars in every garage and consume oil at the US rate, China would need over 80 million barrels of oil a day, exceeding the 74 million barrels per day the world now produces. We are learning that the Western industrial development model is not viable for China, simply because there are not enough resources for it to work. This, in turn, can only add further heat to conflicts over access to dwindling natural resources, particularly as China becomes more economically powerful and capable of out-competing established world powers for critical industrial resources such as oil and metals. We seem to be learning, albeit at a slower pace, that the unreconstructed industrial model has no long-term viability for the West either, although there is an understandable will to hang on to our increasingly precarious advantages, and for the rest of the world to aspire to them. The portents are ominous, and the responses of the 'educated' world appear far from educated or timely.

However, emphasis on merely the many scary facts and figures about our current plight and ongoing trends runs the risk of 'rabbit in the headlights' syndrome: a paralysis of action in the face of seemingly insurmountable challenges. Of course, human progress has been stimulated time and again when we unite and innovate in the face of the apparently impossible, enabling us to overcome such adversities as predation, disease, food scarcity, new climatic zones or wars. Human nature is inherently about seeking solutions, and our progress with various dimensions of the modern environmental challenge offers us crumbs of optimism. We will turn to instances of 'silver linings' fringing the current dark environmental clouds as they pertain to salmonids, as we move into Part 3 of this book, but will now touch briefly on some of the solutions being put in place at a wider geographical scale.

9.10 Solutions

We have the resources to tackle at least some of the most pressing sustainability problems. For example, it would cost about US$200 billion to supply clean drinking water and sanitation for every village, town and city on the planet. This is roughly the same amount of money that is spent on advertising in the US every year. Also, the UK Government's 2006 *Stern Review on the Economics of Climate Change* concluded that the benefits of strong, early action on climate change (estimated at 1% of global GDP) considerably outweigh the likely costs arising from its worst effects, and that failure to do so could risk global GDP being suppressed by up to 20% of what it otherwise might be. The *Stern Review* states that climate change is the greatest and widest-ranging market failure ever seen, presenting a unique challenge for economics as the worst effects, if they are to manifest, may be irreversible. The resources are therefore there, as is a clear self-interest case for averting serious consequences. Whilst one could argue that humans have, for the time being at least, lost that sense of survival that saw us climb to the top of the evolutionary tree, we are now informed by science, economics and, let's hope, nascent political urgency born of growing awareness of the trajectory of current policies.

To date, our collective response to the pressing challenge of sustainability has been slow, but it is evident and steady and we can take heart from at least some technology trends. For example, petroleum accounted for about 2% of total energy supply in 1902 but was already expanding quickly in niche markets. By comparison, renewable energy accounted for about 2% of total generation in 2002, with wind and solar energy markets doubling in size every three years. Renewable energy production and use is soaring worldwide, with double-digit growth in the last decade and signs of market acceleration, such as Worldwatch Institute assessments that, at 40% and 25% respectively in 2008, sales of solar energy equipment have surpassed wind power generation to become the world's fastest growing energy source. In the UK, for example, wind power generation plant grew by 32% in 2006 to reach an installed capacity of 2 gigawatts, almost tripling in the following five years to reach an installed capacity of 5.9 gigawatts. US demand for solar energy roughly doubled during 2010, and was projected to double again during 2011. These levels of growth are on the same scale as the take-off of the internet or of mobile phones, with manufacturers able to scale up production and drive down costs. We seem to be seeing the turning of the technological tide, notwithstanding concerns about some aspects of renewable energy production, particularly its piecemeal delivery rather than any real semblance of an overall renewable energy strategy.

Other natural resources provide mechanisms for peace making, equitable sharing and co-operative governance to achieve a common good. For example, if India, which shares the River Ganges with Bangladesh, were to use all the water that it wants, the Ganges might not even reach Bangladesh during the dry season. Fortunately, a treaty has secured agreements on the allocation of water to Bangladesh. In southern Africa, agreements governing the sharing of water in trans-boundary rivers have provided a means to create, or maintain, peaceful relationships between sometimes war-torn neighbouring states. Other significant international agreements to secure common, longer-term interests include international conventions such as CITES (the Convention on International Trade in

Endangered species) and Ramsar (providing protection for wetlands of international importance), in addition to a raft of statutory EU environmental directives, and nascent carbon markets to address the emerging global challenge of climate change. Although global and EU targets respectively to slow down and to halt the loss of biodiversity by 2010 were not met, concerted global negotiations continue with new targets being set in response to mounting global concerns about the consequences of species loss. The IPCC, discussed previously, is an exemplar of international co-operation to assess information relevant to understanding of risks stemming from human-induced climate change.

We have the capacities for agreement, scientific understanding and economic reformation to make a progressive transition to a sustainable path of development. It may be difficult, but then we have to reflect that it seemed impossible only thirty years ago. Forty years ago, we were barely aware of needing a new pathway at all.

9.11 A new pathway of development

Accusations from business leaders that the 'green lobby' is calling for us to go back to the Stone Age were, as recently as the 1960s through to the 1980s, alarmingly common. As we have grown in societal literacy about sustainability and the lines of cause and effect between ecosystem destruction and impacts for people, businesses and societal wellbeing, these reactionary cries have become less frequent, albeit often still present and veiled in different terms. We have, indeed, seen a sea change in attitude towards broader public recognition of the long-term benefits of protecting the environment as a key asset underpinning current and future human interests, though the extent to which this has exerted meaningful political pressure on governments to act remains moot, leaving the future of our fish and ourselves very much in doubt.

In reality, sustainable progress is far from a regressive step. The reality is that we simply cannot afford to abandon development. Development is not only a 'good thing', allied to the human instinct for progress, but also absolutely essential if all amongst the mass of humanity are to meet their needs and have a reasonable chance of realising their potential. However, what has to change is the way that we frame that development.

The pathway of development upon which we have been embarked over millennia, and increasingly so during the past two and more centuries of industrialisation, has treated the natural world as an inexhaustible supply of resources and services to be mined for our own short-term satisfaction and economic advantage. But we have also grown immeasurably in understanding and self-awareness, such that we can now better inform our choices of technology, resource use including sourcing and fate at end of product life, governance, and so on, with a knowledge about wider consequences over space, time and for the many others who share this limited Earth ecosystem, now and into the future. *Homo sapiens* can only earn the title 'sapient', and the opportunity to live fulfilled lives into the future, if we unleash that capacity in our quest for a more sustainable and equitable pathway of development.

We have spent some pages exploring this breadth of impacts on the global environment, which sets an important context for the scale of change in the world within which local pressures on game fishes are exerted. These broad human impacts on the environment are no longer regarded as of peripheral concern. Some argue we have transitioned from the Holocene epoch into the Anthropocene, the age of the humans, marking an epoch in which human activities are becoming one of the most significant impacts on the functioning of Earth. The cumulative activities of the large and still growing population, with increasing demands and competition for global resources, radically alter the environment at all scales from the local to the global.

Already, it is clear that the prognosis of 'business as usual' for the planet's ecosystems, and the human potential that they support, tells us that yesterday's assumptions, notwithstanding all the advantages they have bequeathed upon the modern world, are inherently too narrow in perspective to account for their wider-scale and longer-term consequences.

We must not stop developing, but learn to do so more intelligently. We have to do so in ways that treasure and use wisely the ecosystems – their integrity, fish and other characteristic living organisms, and the functions that they perform – as resources that are not merely important for, but essential to, our own long-term security and potential.

CHAPTER 10

A brief unnatural history of the British game fishes

T he fortunes of salmon, trout, charr and grayling have not been universally impaired by the agency of human advancement. Indeed, some have spread dramatically in range. However, environmental pressures exerted by humanity have had many dramatic negative implications for these charismatic fishes.

10.1 British game fishes abroad

As humans became increasingly mobile over the centuries, they took with them a wide range of other species. Some, like rats, cockroaches and a host of human diseases, have been accidental and generally unfortunate introductions. Others, such as maize, rhododendrons, cattle, dogs and pigs, have been deliberately spread for agricultural use, or for the pleasure or familiarity they provide.

A dramatic example of this is found in the 'acclimatisation societies' created during periods of European empire building, established for the expressed purpose of 'enriching' the flora and fauna of newly colonised regions with familiar species. The first such recorded society was *La Societé Zoologique d'Acclimatation*, founded in Paris in 1854. However, similar societies spread rapidly and were particularly prolific and active in European colonies in the Americas and Australasia. In many instances, they existed both as societies for the study of natural history, as well as to 'improve' the success rate of introduced species. The appeal of acclimatisation societies in colonies, particularly in New Zealand, stemmed from the widespread belief that the local fauna was in some way deficient or impoverished, fuelled also by the nostalgia of colonials for familiar species and the desire for the introduction of organisms of commercial and/or sporting value. Familiar fish species were part of this package of introductions.

Invasions by non-native species of plants, animals and microbes have often been ecologically disastrous. This is graphically shown in Australia where, due to

Fig 10.1 The brown trout has been widely introduced right around the world.

the vulnerability of its native biota, feral rabbits, foxes, cats, prickly pear cactuses, willow trees and many other invasive species have degraded locally adapted flora and fauna, exacerbated soil erosion, driven endemic species to scarcity or extinction, and otherwise upset the fine balance of local ecosystems, frequently with devastating consequences. In South Africa, alien invasive trees continue to spread rapidly, displacing native species that are generally well adapted to the arid landscape and so accounting for significant volumes of water loss in already arid regions.

Around the world, the introduction of alien fish species beyond the native range and ecosystems with which they have evolved – carp throughout the world; topmouth gudgeon in the UK; tilapia outside their native ranges in Africa and especially across India; the large filter-feeding silver carp in the USA, and many more besides – has radically changed the nature and quality of aquatic ecosystems and water resources.

Similar global damage from alien and invasive crayfish, flowering plants, trees, flatworms, microbes, crustaceans, parasites such as the salmon fluke *Gyrodactylus salaris*, and a host of other groups of organisms, whether wilfully or inadvertently introduced, have created severe impacts, frequently with substantial consequences for economic and wider human wellbeing.

Some species of British game fishes have benefited from being introduced well beyond their native ranges. Brown trout, for example, are widespread in Himalayan rivers, throughout which they have proliferated, following the introduction of their eggs and fry from Loch Leven stock by officers of the British Empire, eager for familiar sport. In the remote mountain kingdom of Bhutan, brown trout of similar descent flourish in the cold, clear waters of one of the highest lakes in the world. Brown trout were introduced to New Zealand in 1867, thriving in waters similar in character to those of their native range. Also, introduced brown trout have prospered in the upland waters of the Western Ghats chain of mountains rising from the western seaboard of peninsular India, as indeed they do in upland regions of South Africa and Lesotho, widely across North and South America, the Falkland Islands, Australia, Turkey, the Jordan River, Tanzania and Kenya, Japan, Pakistan, Swaziland, Fiji, Papua New Guinea, Morocco, Ethiopia, Cyprus and Nepal.

So too, Atlantic salmon escaping from fish farms have formed viable breeding populations in river systems on the Pacific coast of north America. We will discuss these, and other wider implications of caged salmon farming, later in Part 2 of this book.

Arctic charr have generally not been so widely spread by humans. However, in Siberia, where they are known as 'golets', their introduction to lakes has sometimes threatened endemic species, including the less robust and more localised long-finned charr (*Salvethymus svetovidovi*).

Grayling have been spread for their sporting virtues well beyond their native range, which spans northern Europe from England and France in the west eastwards through to the Ural Mountains of north-western Russia. Grayling are not native to Scotland but are now widespread following their first introduction to the Clyde from Derbyshire in 1855. In subsequent years, populations have been established in Scotland's Rivers Annan, Ayr, Earn, Nith, Tay, Teviot and Tweed, where they thrive and have spread into tributary river systems. Grayling are now also firmly established following introduction into many Welsh rivers, including the extensive Severn, Wye and Dee systems. Grayling have also become established following introductions into south-west England, thriving locally but with a patchy distribution in two rivers in Devon and Cornwall: the cool, clean flows of the River Exe and Tamar systems.

Viewed through the narrow blinkers of the success of British species, or the enjoyment of expatriate anglers, the pervasion of familiar species into new waters may be regarded as beneficial. However, we have to understand the impacts of these introductions from broader perspectives, and knowing the massive harm inflicted on water resources and ecosystems across the world by the incautious spread of fish species such as carp as well as other invasive animals and plants. There is certainly scientific evidence of predation and competition by brown trout on the balance of ecosystems in New Zealand and Australia, including impacts on the distribution, and sometimes local extinction, of some small native galaxid fishes, frogs, crayfish and other invertebrate fauna. Equally, competition for optimal spawning gravels, juvenile habitat for parr and other important niches exploited by Atlantic salmon colonising the Pacific coast of America, are all likely to have negative consequences for the success of native Pacific salmon (Chinook, coho, sockeye, pink, chum, cherry and Amago salmon, all of the genus *Onchyrhynchus*). Furthermore, introductions of grayling to those Swedish rivers in which they are not native, and where they have not cohabited with native fishes, has been found to have adverse consequences for native brown trout populations.

10.2 Foreign salmonids in British waters

A number of non-native trout species and their hybrids have been introduced into British waters. Some have formed self-sustaining populations, though others

Fig 10.2 The rainbow trout is an invasive and still widely stocked species in British waters, introduced from the Pacific seaboard of America.

have not. Rainbow trout (*Onchyrhynchus mykiss*) have been especially successful and are now spread across the British Isles as wild, self-sustaining populations, as popular stocked fish in still and flowing waters, and as escapees from their extensive production in aquaculture.

Other less common alien introductions to British waters include brook trout, golden trout and cutthroat trout (all from North America), varieties such as blue trout (rainbow trout lacking the red stripe) and various hybrids including the tiger trout (a cross between brook and brown trout).

10.3 Controlling the spread of alien species

The historic desire for expatriates and émigrés to take familiar species and associated lifestyles with them to new lands found its zenith in the acclimatisation societies of the colonial era, the many and diverse negative consequences of which still harm ecosystems and human interests centuries later. But through both deliberate and accidental introductions of species through horticultural, agricultural, aquacultural, food and associated trades, in addition to the inevitable stowaways of modern globalised transport of people and goods, we are learning about the dangers of invasion by alien species freed of the checks and balances of the ecosystems with which they have evolved over millions of years.

We will turn to the consequences of the spread of alien species later in this book. It suffices for current purposes to note that the spread of salmon and trout, both to and from the British Isles, has the potential to seriously disrupt the ecosystems into which they are introduced.

CHAPTER 11

Net results

The harvesting of fishes from seas, estuaries, rivers and still waters is a craft as old as humanity. Fish form a staple part of the diet and a key source of protein for many human communities across the world, both today and throughout our long history since emerging beside the lakes of Africa's Rift Valley some 150,000 years ago.

However, the explosive growth of the human population, allied with our increasing technological capabilities, has placed ever-greater pressures upon aquatic and many other natural resources. The fishes of the salmon family are a prized target species. However, they have proved to be vulnerable to over-exploitation by commercial fishing interests, compounding the many wider pressures on their survival.

11.1 Direct exploitation of native game fish stocks

Trout were once widespread in British rivers and were harvested locally as an excellent high-protein food supply over the centuries. Later, wild brown trout also found their way onto national markets, before commercial fish farming became established to satisfy the demands of fishmongers and catering outlets. It is also fair to say that angling for wild trout made substantial inroads into native stocks, so much so that hatchery-reared brown trout have been stocked into rivers to augment supplies for anglers for well over a hundred years, inevitably resulting in the further reduction of truly wild, post-ice-age gene banks of brown trout in UK waters. However, quite how deep this impact on genetic integrity is remains open to conjecture until such time as we are armed with credible data from detailed genetic analysis. Nevertheless, as discussed elsewhere in this book, exploitation is only one of the issues affecting the decline of wild trout. Probably just as culpable, although still of human origins, is the degradation of the freshwater environment upon which these remarkable fish depend.

Beautifully adapted as they are to harvesting the productivity of the sea before returning to breed in fresh waters, there survive various tales of Atlantic salmon

Fig 11.1 A salmon netting boat takes to sea.

once being so common that various poor people had written into their contracts that they would not be fed with the fish too frequently. However, it is somewhat harder to come by evidence to substantiate these oft-repeated 'urban myths'. It is said, for example, that Atlantic salmon were once so abundant many years ago in the Gave d'Oloron, which remains one of France's main salmon rivers, that peasants insisted on a limit for being fed the fish for lunch of three times a week. The tale of London apprentices having in their contracts that they would not be fed excessive salmon from the Thames is perhaps the most widely cited tale of this type, but supporting documentary evidence is slight, comprising a passing mention in the *Itinerary of John Leland* written on or about the years 1535–1543 by the Antiquary John Leland (1503–1552).

Fig 11.2 A set net, a highly effective salmon killer, operating south of Montrose (image courtesy of Andrew Graham-Stewart).

Fig 11.3 Open sea capture fisheries were once at truly sustainable levels.

Across the Atlantic, common folklore in New England tells of Atlantic salmon populations so abundant that early colonists could 'walk across the backs of the fish' as they ran up the rivers to spawn, with tales that people became so bored of eating them that a law was passed requiring poor servants and labourers to be fed the fish no more than twice a week. Indeed, at the time of European colonisation, the Atlantic salmon was reportedly found in every river not barred by impassable falls from Labrador possibly as far south as the Hudson River. However, analysis of pre-colonial domestic debris left behind by Native Americans reveals that, unlike those on the Pacific seaboard, there was an almost complete lack of salmon bones in New England despite there being an abundance of bones from numerous other fish species. Study of the literature reporting these rich salmon stocks revealed that most was written or compiled well after the colonial period, and so subject to bias and misinterpretation. As long ago as 1867, the *London Field* magazine is reported to have offered a reward of five pounds to anyone producing one of the oft-quoted agreements, but the reward was withdrawn unclaimed a year later.

We all know, of course, that people throughout the ages have assumed that fishing and fish stocks in earlier times were better than they are today, and that the rose-tinted nostalgia for some mythical yesteryear can blind us to objective analysis. This seems to be the case for the Atlantic salmon of the USA, and very probably on the European side of the Atlantic as well.

As we have already seen, Arctic charr fisheries were once prosecuted in the UK, especially in Windermere, where the fish were often potted to aid their preservation in the days before freezing became the norm. However, interestingly, exploitation by humans was not the only predatory influence on Windermere's charr stocks, which declined markedly in the first third of the twentieth century. At this time, pike and perch populations were on the increase, and it was not until the Freshwater Biological Association began to control these predators that charr made a significant comeback, at least for a while. Alas, water quality issues now plague Windermere, with sudden pH spikes as high as 11. Combined with climate change and nutrient enrichment issues, it is doubtful whether Windermere charr will return to their historic numbers in the foreseeable future.

Grayling have only ever been exploited at a very local level as a food resource, even though many people from older generations in some areas of the country would prefer a brace of grayling to a brace of trout for supper. But grayling have been historically persecuted, at least during the past hundred years or so, as a perceived competitor to brown trout; a more valued species in angling terms. For decades, fishery managers caught and killed large numbers of grayling in the misconceived belief that this would benefit the trout in terms of increased food and habitat resources. Likewise, many game anglers who caught grayling while fishing for trout would often throw them in the nearest bush or, at best, take them home for pet food. However, as we have seen in Part 1 of this book, grayling are not strongly competitive, occupying a separate niche in the freshwater environment. It is only recently that perceptions have changed, helped enormously by publicity from such organisations as the Grayling Society, and grayling have at last begun to be accorded some of the respect that they deserve.

If harvesting of wild fish is controlled such that the take does not exceed natural regeneration rates, populations of these game fishes represent a valuable and also a potentially sustainable source. However, if taken in quantities beyond this renewal rate, or if habitat and/or other environmental conditions compromise reproductive success, population decline is inevitable, and this valuable resource will be degraded or lost.

11.2 Plunder on the high seas

High-seas netting for salmon is, funnily enough, a fairly recent affair, although coastal netting has taken place for centuries. Scottish fishermen opened a drift net fishery in the early 1960s but, following a recommendation in the 1963 *Hunter Committee Report*, the fishery was promptly closed without compensation for

Fig 11.4 A netted Atlantic salmon.

netsmen. In England, the large drift net fishery off the north-east coast, in which 64 licence holders had killed 27,685 fish in 2002, was significantly reduced in 2003 in a buy-out partnership between the government and private fishery interests. However, 15 netsmen were still in business in 2010 and, together with the increasingly effective fixed nets on the Northumberland and Yorkshire coasts (discussed in 11.3 below), a total of 19,982 salmon were killed in the north-east fishery, bringing into question English salmon netting management policies.

By far the most important, and the most damaging, high-seas fisheries were occurring in the 1980s and 1990s off the coasts of the Faeroe Islands and West Greenland. These were truly mixed-stock fisheries, for this area is the most important feeding ground for salmon originating in most of the Northern hemisphere's countries with runs of Atlantic salmon. The high-seas fisheries therefore did untold damage, not only to national stocks but also to those from individual rivers. This method of indiscriminate capture bore no relationship with whether natal river systems held healthy or struggling salmon populations. The efficient management of salmon stocks was therefore gravely threatened, and it soon became clear that action was required if irreversible damage was not to be inflicted on the whole northern hemisphere stock.

The North Atlantic Salmon Conservation Organisation (NASCO) was formed in 1989 as a result of a resolution taken at an Atlantic Salmon Trust conference in Edinburgh. Its original brief was to achieve catch quotas for the Greenland and Faeroese fisheries, and so give national governments greater opportunities to manage their stocks better and also reap the socioeconomic rewards associated with both angling and net fisheries. However, partly instrumental in this initiative was the compensation fund set up for Greenland and Faeroese fishermen by the North Atlantic Salmon Fund (NASF), a partnership of NGOs from around the northern hemisphere led by the redoubtable Icelander Orri Vigfusson. Political agreements and commercial compensation packages therefore formed a formidable partnership in the conservation of salmon stocks on their marine feeding grounds, although it is fair to say that there was precious little co-operation between the two initiatives and, as so often happens in fisheries management, the combined success was in spite of co-ordination, rather than because of it!

The result is that the Faeroese have retained their right to fish for salmon on the high seas but have not exercised that right since 2000. Meanwhile, Greenland agreed a subsistence tonnage for home consumption: just ten tonnes in 2010, although this crept up to around forty tonnes by the end of the season, but is still well short of the 2,000+ tonnes caught annually in the past.

However, with NASCO monitoring teams taking samples from salmon caught from the Greenland fishery, and genetic marking of fish from individual river systems really gaining momentum across the Northern hemisphere, we are beginning to gain invaluable information as to where different national stocks

Fig 11.5 Beach netting, referred to as 'net and coble' in Scotland, for salmon and sea trout (image courtesy of Andrew Graham-Stewart).

feed. Thus, together with initiatives such as the EU-funded 'Salmon at Sea' (SALSEA) Project, which netted post-smolts and adult Atlantic salmon at sea to establish data on migration routes, we are beginning to learn more about the marine cycle of the salmon's life, enabling us to inform future management policies with far more accurate and pertinent scientific information.

11.3 Coastal netting

With the closure of commercial netting on the high seas, the focus has returned to the coastal fisheries of northern hemisphere countries. If the fish had at last been spared off the Greenland and Faeroese coasts, we were all honour bound to ensure that as many as practically possible were able to run their natal rivers, where local management decisions could be taken to ensure adequate spawning escapement.

The Republic of Ireland and the north-east England drift-net fisheries were relatively easy to close and reduce respectively, albeit over extended timescales and entailing much negotiation. This was achieved principally because they were both licensed and controlled by governments. In Scotland, however, rights to fish are heritable, and therefore the government has less instant control over management options. However, effective legislation is available if the political will can be found to apply it.

Data from the Environment Agency's *Fisheries Statistics Report 2008* tell us an interesting story about trends. In England and Wales in 2010, 22,364 salmon and

29,071 sea trout were killed by all netsmen with the majority (88%) of these taken in the north-east drift nets and coastal T&J nets (fixed engines placed at right angles to the beach). While the north-east drift nets were substantially reduced by the 2003 partial buy-out, the T&J nets are catching an increasing number of salmon and sea trout leading to a north-east coast doubling of salmon catch from 2008. This is a concerning trend at a time when both salmon and sea trout stocks are under pressure throughout the UK, and one can only hope that the research currently being undertaken into sea trout life-cycles will inform better management policies in future.

The Scottish coastal fisheries still take significant numbers of fish. In 2010, 15,577 salmon and 2,360 sea trout were caught here by fixed engines, significantly impacting mixed stocks. Like the Greenland and Faeroese fisheries, these coastal netting stations are indiscriminate in the fish they catch, unable to differentiate whether captured salmon were heading back to a natal river with a strong or weak population. With the closure of the Irish drift net fishery, only Scotland, England, Northern Ireland and particularly Norway have coastal mixed-stock fisheries remaining; all other countries with salmon runs have closed these damaging netting stations down, and even the English nets will phase out over time, as current licenses will not be reissued to new applicants, even to family members, when fishermen retire or leave the industry. Needless to say, therefore, the NGO community has its sights set on closing all other coastal fisheries as quickly as possible. Only then will management truly revert to the catchment scale, where more efficient decisions can be taken according to local stock status.

11.4 Estuarine netting

In Scotland, 11,738 salmon and grilse and 8,663 sea trout were caught in estuarine and in-river net and coble fisheries (cobles are sweep nets shot in a wide circle from beach or bank using boats) in 2010. However, due mainly to local buy-out initiatives and, hence, a greatly reduced netting effort in estuaries, this is a mere fraction of historical levels, albeit still significant. Also, of course, these netting stations come under the management policies of individual District Salmon Fishery Boards, and so can be regulated, along with rod and line fisheries, according to local stock status.

In England and Wales, again, buy-out schemes have greatly reduced netting effort, and only a small proportion of the English and Welsh net catch now occurs in estuaries. We have probably reached a satisfactory level of exploitation with estuary netting, because it would be sad to lose completely the heritage value of salmon netting, which, after all, sustained our forebears for centuries. This, together with the body tagging scheme discussed in Part 1 of this book, is believed to substantially reduce the legal and illegal taking of sea trout, better to align it with sustainable levels of exploitation.

However, the final word on all commercial netting, whether on the seas, in coastal waters or in estuaries, is that all fish have to die to contribute to local socioeconomics, and this inevitably has management implications. Rod and line fisheries can still contribute strongly to local economies while exercising conservation restraint by adopting a 'catch and release' ethos, which, with the high survival rate achieved in most fisheries, adds greatly to spawning escapement and the future wellbeing of the stock.

11.5 Rod and line

Rod and line angling for game fishes remains extremely popular throughout the British Isles, despite generally poorer runs of migratory fish than a generation ago. However, as with coarse fishing, there has been something of a move to stocked stillwaters, which, while not guaranteeing a catch of fish, at least gives a much better chance of attracting one's quarry than attempting to lure a wild specimen from a more natural and unpredictable habitat.

Some 36,461 licenses were sold in 2010 by the Environment Agency to fish for salmon and sea trout in England and Wales, which accounted for 15,527 salmon and 30,204 sea trout, of which 58% and 62%, respectively, were released. However, early-season fishing is undoubtedly affected by the Spring Byelaws, brought in by the Environment Agency in 1998, prohibiting the killing of salmon before 16th June. A decade after their introduction, a 2008 review of the Spring Byelaw found that, despite widespread abhorrence of compulsory 'catch and release' at their inception, anglers' attitudes to conservation had changed so markedly in the interim that there was a high degree of support to keep the initiative alive for a further ten years. Even those rivers with an arguable surplus of early-running salmon, such as the Tyne, voted to keep to the byelaw, mainly over concerns that the river would be inundated by anglers trying to catch a springer 'for the pot', should this river alone be granted dispensation.

In Scotland, angling numbers have undoubtedly been affected by foreign fishing destinations in recent years. Russia, Iceland and Alaska now lure the well-heeled angler, with their virtual guarantees of heavy catches, and exotic marine fly fishing for bonefish, tarpon and similar sporting quarry has also proved hugely attractive to game fishers. Further south, the sea trout and wild brown trout fishing of South America has become an equal draw, despite these fishes being introduced to those southern hemisphere waters, and many more affluent anglers now make at least an annual pilgrimage to one or more of these regions. Even closer to home, marine fly fishing for bass, mullet, mackerel and pollack is one of the fastest growing sectors of the sport, especially in summer when hot weather and low flows render traditional angling for freshwater game fishes extremely difficult in most rivers.

Scottish rivers, too, have instigated conservation measures in the past decade, though with a different, catchment-based management system centred on individual river systems rather than by national legislation. So, for example, the Tweed, Scotland's most prolific salmon river, brought in a compulsory 'catch and release' regime in 2011 for all salmon caught up until 30th June so as to protect the increasingly fragile early running, or 'spring', component of the river's stocks. Scotland's River Dee is now totally 'catch and release' and, interestingly, this has attracted a new client base to many fisheries, as traditional anglers have moved away from compulsory conservation to be replaced by those more attuned to modern, more sustainable ways of thinking.

Scottish anglers killed 32,712 salmon in 2010, but another 77,784 were caught and returned alive to the water, constituting a 70% 'catch and release' rate, up from 62% in 2008. Also, 7,843 sea trout were killed in 2010, but a further 18,961 were caught and returned (a rate of 66%, rising from 56% in 2008). This shows just how far anglers' attitudes to conservation have come in all parts of the country, helping to bring the UK into line with Russia, Alaska and many rivers in Iceland (as well as UK coarse fishing), where 'catch and release' is the norm and very few anglers worry about it anymore.

11.6 Illegal harvest

While anglers do their bit for conservation, poachers still ply their trade to an irksome degree. By the very nature of the activity, it is impossible to put accurate figures on the illegal harvest of British game fishes, which comprise mainly Atlantic salmon and sea trout. However, there is no question that poaching is not as rife today as it once was, due primarily to the deleted runs of wild fish in many areas and the cheap price of farmed fish. However, illegal harvesting does still exist and can cause specific problems locally which, with the diminishing resources available to statutory authorities for enforcement, is often difficult to police. This has become somewhat easier in England and Wales since the beginning of 2009 with the passing of a national byelaw prohibiting the sale of rod-caught fish and, most importantly, introducing a carcass-tagging system for commercially-caught salmon and sea trout. Any fish on the trader's slab or in the restaurant's kitchen which does not carry a unique identification tag is considered illegal, and so the enforcement can now be more efficiently focused at the consumer end, rather than being spent on long and risky surveillance missions in the faint hope of catching poachers 'red handed' on the river bank.

In Ireland, both north and south of the border, tagging has been successfully established for some time, and the sale of rod-caught fish was banned in Scotland under the Aquaculture and Fisheries Act 2007. However, the Scottish Government did not impose a carcass tagging scheme, and so one of the major recommendations from a fisheries working group operating throughout 2009 was for this to be implemented as soon as possible, so closing a potential ready market for fish caught illegally anywhere in the British Isles.

However, as has been mentioned elsewhere in this book, other environmental and man-made factors come into play with poaching. For example, local fisheries interests have fought for several years to have the fish pass on the Tees Barrage made more efficient in allowing salmon and sea trout to run through it. The main consequence of a hard-to-pass barrier such as this is that fish stack up below it in unnaturally large shoals and become easy prey for both mammals and humans. So, the thriving population of Tees estuary seals has had a glut of fish in recent years, and a steady trickle of Environment Agency prosecutions for human poachers suggests that many fish have ended their days being consumed illicitly by people.

Likewise, if a river is closed to angling for any reason, such as occurred in the Irish Republic in 2007 as a political sop to the netsmen for the closure of their drift net fishery, the 'eyes and ears' of the angler are lost as an informal enforcement tool. Poachers know that, and so the very act that was supposed to conserve fish stocks ends up perversely having the opposite effect. Hence, organisations such as the Salmon & Trout Association have lobbied for river fisheries to be kept open for angling if at all possible, even if this means imposing local compulsory 'catch and release' restrictions.

If anglers are on the riverbank, they act as deterrents to poachers, and they attract continued investment into the river by fishery owners, while 'catch and release' optimises spawning escapement. This represents a 'win–win' situation.

11.7 In conclusion

Cumulative pressures on a diminishing wild population, threatened by additional environmental pressures that we will explore in the remainder of Part 2 of this book, are placing huge stresses on native salmonid fishes. The impacts of commercial capture of Atlantic salmon and sea trout have changed significantly over time, but 'silver linings' are apparent in the way in which policy, incentives, commercial practice and changing attitudes amongst a more environmentally aware angling community are palpably adapting to these pressures.

Like the infamous curate's egg, capture of game fishes of the salmon family, particularly migratory species, is 'good and bad in parts'. We need economic returns in order to support conservation of stocks for anything other than altruistic reasons, yet there are sustainable limits to this that have to be respected, informed by sound science, and imposed rigorously through enforcement, economic instruments and voluntary codes.

The role of 'catch and release' recreational fisheries, which are based on the value of living and not dead fish that are then returned to river systems to pass their genes on to the next generation, is important for sustaining local economies and traditions.

Chapter 12

Muddying the waters

Landscapes and waterscapes have varied constantly throughout evolutionary history. The British landscape has changed dramatically, both through and following the retreat of glaciers under successive ice ages, and through the rising of sea levels, which cut off land connections formerly linking us to continental Europe. Coastal erosion and deposition, and the tipping of the geological plate upon which our small islands lie, conspire to continuously reshape the shoreline and its diverse cliffs, sand dunes, saltmarshes and strandlines. Inland water, in the form of rainfall, ice and river flows, has carved familiar landscapes, occasionally altering their points of discharges to the sea as they erode and break through land barriers. The British landscape with which we are so familiar is, viewed from a perspective longer than our limited life spans, in constant flux with no single long-term period of complete stability, including the character of its diverse wetlands, rivers and standing waters.

12.1 The human landscape

Change has been an accelerating constant since humans started to exert their own dramatic pressures over and above the forces constantly shaping the natural world. In pre-Roman Britain, much of the wildwoods of mature trees, representing climax vegetation, still covered the mainland. Rivers flowed through them in multiple, braided channels of great complexity, with woody debris, exacerbated by the activities of beavers, forming lagoons in some limbs whilst fast flows scoured clean gravels in others. River habitats would have been in constant flux, spates forming clearings rich in wetland flora and supporting associated grazing animals, becoming progressively enclosed once again as new trees grew up in a succession. Our moist Atlantic climate, warm for its latitude because of the Gulf Stream's influence, enabled the formation of a wealth of wetlands across these small islands, collectively forming a great sponge that stored and smoothed flows down rivers, purifying water and supporting diverse game and other wildlife.

Fig 12.1 A well-vegetated, little-managed river provides diverse habitats for many species and their different life stages.

This matters for fishes of the salmon and other families and for all aquatic life, for it is within this habitat-rich and dynamic environment that they evolved. However, the shaping of the modern world has since exerted all manner of pressures on nature. Successive periods of human history, from the origins of settled agriculture to the formation of cities, the forces of the Industrial Revolution and a range of other major periods of transition, have made the face of the landscape of Britain, and much of the rest of the planet's land masses, unrecognisable from former times.

As we have seen in the previous chapter, the spread of members of the salmon family and other fishes from, and into, British waters, not to mention a wide range of other alien and often invasive species of plants and animals, has had dramatic implications for their character, ecosystems and overall vitality. Would that the translocation and exploitation of game fishes were the only problems faced by native stocks! The making of the modern world has wrought many and substantial changes to the characteristics of the aquatic systems upon which all species – fish and humanity alike – ultimately depend. Furthermore, fundamental changes in the balance of gases in the atmosphere have had, and will increasingly have, a profound influence on the global climate exacerbating other environmental impacts. Game fish populations across the British Isles are, indeed, under grave threat from a variety of pressures.

12.2 Changing landscapes

Land management across substantial swathes of the countryside, exacerbated by urban and industrial sprawl, has changed many of the characteristics of our rivers. The widespread loss of wetland systems through the millennia has radically altered landscape hydrology, generally making rivers increasingly flashy and vulnerable to both extreme high and low flows. Meanwhile, erosion through excessive grazing by stock animals as well as tillage for arable farming, generally without adequate buffers between farmland and river edges, has substantially changed the bed materials of many British rivers.

Siltation within rivers is a major problem right across the country, as indeed much of the densely inhabited world, and particularly so for those reaches in which game fishes traditionally excavate their redds. Not only does excessive silt input to rivers block the pores in coarse gravels, preventing the flushing of buried eggs by cool oxygenated waters, but the river beds may themselves become compacted, such that redd digging becomes difficult or impossible. Water within the sediment may also become depleted of oxygen and, therefore, unsuitable as habitat both for spawning and for the production of adequate insect food for fry, parr, smolts and adult fish.

Excessive inputs of nutrients, particularly of phosphorus and nitrogen from agricultural, sewage and industrial sources, as well as remobilisation from sediment inputs to river systems, are also potentially problematic. The impact of eutrophication, wherein production of algae and other plants is boosted by high levels of nutrients, includes the blocking up of river sediments, changes in oxygen levels and invertebrate communities within rivers and even, in some circumstances, direct toxic effects upon fish. In still waters, deoxygenation of deeper levels of lake systems can result from the decomposition of planktonic algae, which may also impede light penetration, suppressing the growth of water plants on the lake bed and margins, often with significant changes in the balance of the entire ecosystem. This can be particularly problematic for glacial relic species such as Arctic charr and the whitefishes, as well as other game and coarse fish species exploiting the cooler depths of still waters.

Fig 12.2 Urbanisation severely affects habitat and many other facets of salmonid rivers.

Fig 12.3 Excretion and bankside destruction by cattle cause damage at scales that would see factories closed down.

Fig 12.4 Farmyard run-off into watercourses can be highly polluting.

Fig 12.5 Algae growth following excessive nutrient run-off.

All of these impacts clearly create stresses for Britain's game fishes, affecting their overall health and, hence, their vulnerability to other factors such as predation, pollution, parasites and diseases, rising temperature and declining river flows.

12.3 Poisoning the well

Pollution of fresh, brackish and saline waters occurs from a variety of sources. Throughout the last quarter of the twentieth century, a great deal of investment was expended in the developed world to deal with point sources of pollution. These include piped discharges from sewage treatment works, industrial complexes, private residential estates, concentrated run-off from major road systems and so forth, which are relatively easy to identify, characterise and treat using various technologies.

However, as overall inputs of pollutants from identifiable point sources began to decline, the magnitude of diffuse pollutant loads arising from a variety of agricultural, residential and industrial sources, including complex pollutant pathways such as aerial deposition, began to be revealed. By their very nature, diffuse sources of pollution are far more difficult to identify, characterise and control. The battleground has therefore shifted somewhat in the way we control human pollution of our waterways.

Fig 12.6 Effluent from point sources has come under increasingly stringent control, but can still be problematic for rivers and lakes.

12.4 An alchemist's brew

Nature also has her own diverse arsenal of 'chemical warfare' agents, evolved by organisms as chemical defences against their grazers, predators or competitors. For example, the hop, a widespread European climbing plant cultivated for the brewing of beer and used in herbalism to promote sleep, produces hormonal bio-chemicals powerful enough to perturb the menstrual cycles of women picking the crop by hand. The world is full of plant, microbial and animal-based toxins, hor-monal substances and other chemicals, some of which have been commercialised as human and veterinary drugs and other chemical agents, such as the natural plant-derived pesticide pyrethrum. Many more of these natural substances remain unexploited or unknown but, where they are released in relatively concen-trated forms into the environment, disruption of ecosystems is to be expected. This was graphically illustrated recently by the discovery that natural mustard oils released into the water during the harvesting of water cress in the headwaters of chalk streams was having a severe impact on the *Gammarus* freshwater shrimp populations for a significant distance downstream.

Increasing volumes of man-made chemicals now enter the pollution stream in addition to the generally simpler chemicals (such as organic matter and nutrient substances in sewage and the effluent of food-processing industries) poured into the natural world by society. This has been particularly the case as synthetic chemistry has accelerated in the post-Second World War era. Manufacture and incautious use and disposal of man-made organic chemicals is likely to result in unpredictable biological consequences, due to cellular enzyme pathways not hav-ing encountered them and therefore not needing to develop detoxification mech-anisms over evolutionary timescales. Whilst larger molecules, such as durable plastics, are unlikely to penetrate cellular membranes, smaller organic molecules are likely to be actively or passively absorbed into cells as they are inherently similar to the diverse organic matter metabolised by organisms. In the absence of enzyme pathways to break them down, there is a risk that they will not only

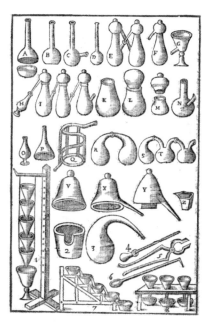

Fig 12.7 Modern chemistry has brought many benefits to humanity but also many unintended consequences.

systematically increase in concentration but also exert unforeseen but potentially harmful effects.

Furthermore, the huge and ever-expanding number of man-made and naturally derived substances in human use, in various medicinal, industrial, agricultural, domestic and a host of other forms, do not simply remain where they are used or applied. Entropy, a consequence of the Second Law of Thermodynamics under which matter and energy tend to disperse spontaneously, determines that, for example, pesticides applied to treat a specific animal, kill crop pests in a target field or delivered to treat a human disease, will not remain in the place in which they are applied. They will instead disperse into ecosystems where, in some cases, they may accumulate in the cells of living organisms with largely unforeseeable consequences.

As we have seen in Chapter 9, The making and breaking of the modern world, many human innovations have, regardless of unanticipated longer-term impacts, produced for the lucky minority of us in the developed world, unprecedented levels of public health, longevity and personal wealth. This includes the benefits of synthetic chemistry, with its diversity of plastics, pharmaceuticals, nutritional aids, cleaning substances and other applications that potentially add to our comfort and security. However, because our historic focus has been upon narrowly focused and immediate advantage, we have frequently been blind to the unfortunate potential consequences of our innovations and decisions once those substances are liberated into ecosystems.

Take for example DDT (dichlorodiphenyltrichloroethane), perhaps the best-known and most notorious of a range of organochlorine pesticides developed in the post-war era. DDT was first synthesised in 1939 and used for a wide range of applications, including the control of mosquitoes and lice, and as a prophylactic and treatment chemical in arable and stock farming. Between 1950 and 1980, some 40,000 tonnes of this new 'wonder chemical' were used each year, largely for agricultural purposes. In part, this breadth of applications reflected the broad spectrum of toxicity of the molecule, which was also persistent by virtue of a structure alien to nature and which was therefore not readily degraded by enzyme systems. As we have seen, persistent substances that do not readily break down in living cells and ecosystems are prone to bioaccumulate, with unpredictable effects. Rachel Carson's seminal and world-changing 1962 book, *Silent Spring*, was a wake-up call to the world, a masterful collation of strands of evidence of the detrimental, unpredictable long-term consequences of pesticide bioaccumulation, and titled with an enduring metaphor of a landscape from which our songbirds had been eliminated. DDT and other organochlorine substances, in addition to organophosphorus and other families of synthetic biocides, were found to be accumulating and persisting widely in the environment as an unforeseen consequence of their use in a host of applications.

All of these impacts were significant for the water environment, but perhaps the most direct impact upon game and other fishes has been the incautious use and disposal of synthetic sheep dip chemicals. The synthetic pyrethroid chemical, cypermethrin, gives rise to particular concern. This powerful agent is lethal in minute concentrations to various target invertebrate groups, and on the juvenile stages of salmonids and the olfactory performance of adults, leading to poor spawning activity in some populations. As a treatment or prophylactic for sheep diseases, cypermethrin and other sheep-dip chemicals have clear veterinary value. However, as a water pollutant able to deplete the invertebrate and fish communities of streams and rivers in only minute concentrations, these substances continue to wreak devastation on aquatic ecosystems. Cypermethrin continues so to do despite its sale having been banned in the UK since 2008. This ban was driven largely through the political lobbying of fisheries NGOs such as the Salmon & Trout Association and Atlantic Salmon Trust. Cypermethrin is still used legally in other agriproducts, which can be utilised as sheep dips, and there is a widespread application of the chemical in young forestry plantations. Not only is cyper-methrin directly toxic to target species, it also contaminates sediments and water potentially abstracted for irrigation and potable uses, creating human health risks and a range of other costly and often insidious impacts on environmental, public and economic health.

Fig 12.8 Incautious disposal of sheep dip, and the use of inappropriate chemicals, can cause massive harm to entire river ecosystems.

Chemical contamination by synthetic and persistent chemicals of a variety of types impacts aquatic birds and mammals as well. For example, the UK otter population crashed catastrophically between the 1950s and the 1970s, contracting in range to a few strongholds in south-west England, Wales and Scotland, with predatory birds such as buzzards and sparrowhawks following similar trends. The contribution of certain persistent agrochemicals to this sharp decline, perhaps even near-extinction locally, is demonstrated by the subsequent progressive recovery following bans on the use of these substances. Today, there has been a resurgence of raptor numbers across the British Isles, and traces of otters have been found in every county of England. Campaigning environmental groups with interests in game fishes, as well as wider ecosystems and human health, have been prominent in campaigning against successive forms of sheep dip and other harmful agrochemicals and their irresponsible use.

Another of the alarming and unforeseen consequences coming to light as a consequence of the accumulation of persistent organic pollutants in ecosystems was that of endocrine disruption. This is the phenomenon of a chemical simulating or interfering with hormonal systems. As we have seen in the example of hops, some naturally produced chemicals can exert powerful oestrogen-mimicking effects. However, it is the broad range of synthetic substances that give the greatest cause for concern, as screening of the impacts of various families of these substances, ranging from birth-control pills to pesticides, surfactants and preservatives, reveals that many have hormone-active properties of potential concern. Historic regulatory frameworks, developed to address far simpler toxic and other cause-and-effect impacts, are poorly-suited to these surprising effects and to control their release into the environment from millions of dispersed homes, vehicles and applications in pharmaceutical, horticultural and myriad other sources.

And, of course, endocrine disruption is only one of a number of impacts identified only long after widespread commercialisation and use of these chemicals. Many more forms of long-term and complex impacts surely remain to be revealed, dictating a far more precautionary approach that takes account of potential implications rather than already proven impacts.

12.5 When the well runs dry

The adverse impacts of physical and chemical inputs to watercourses are exacerbated by excessive abstraction of water. As we have seen, we live in a water-stressed world where, even in developed temperate countries, demands upon water resources often outstrip renewable capacity. Throughout the British Isles, but particularly in the more water-scarce south and east of England, there is a great dependence for public water supply upon abstraction of groundwater,

Fig 12.9 Excessive abstraction of water from surface waters and underground can dry up rivers.

which can rob rivers of flows vital for sustaining the needs of their characteristic ecosystems. Elsewhere, except in poorly populated areas on impermeable geologies to the west of the country, abstraction from rivers and other surface waters can be problematic where river flows decline below that necessary to provide migratory triggers, flush sediment from gravels or to support other needs of different stages of game fishes.

Excessive and inequitable human demands for water, well in excess of natural carrying capacity nationally as well as globally, have already been described in some detail in Chapter 9, The making and breaking of the modern world, so, although this book is not primarily about the global human 'footprint' on ecosystems and the international and intergenerational equity issues this raises, the reality revealed by these issues and statistics do at least serve to demonstrate in blunt terms that modern lifestyles are profligate and that our daily habits and escalating demands inevitably have substantial impacts on the ecosystems that we depend upon to underwrite our future needs, including support for valued game fishes. Our historic demands upon water resources, withdrawn in a utilitarian way with little or no regard to the quantity, quality and flows required to keep aquatic ecosystems functioning, is a major contributor to the threats facing game-fish populations.

12.6 Draining the land

The cumulative impact of all these pressures is also exacerbated by the history of over-management of river systems in the British Isles. From the Second World War through until at least the late 1980s, the emphasis of public policy with regard to land use was to increase drainage to maximise the productivity of agricultural land and to achieve self-sufficiency in food. Realising this economic goal had many wider implications for other benefits that the landscape provides to people, including many profound impacts upon river systems.

There was, for example, a massive depletion of the national wetland resource, as marginal land was drained for food production. This resulted in the loss of the many beneficial services provided by wetlands, including the buffering of flows, purification of water, changes to valued landscapes, support for scarce and valued wildlife such as species of wading birds, and the recruitment of juvenile fish. Gravels were removed from river channels, which not only robbed them of their capacity to support game fish redds, but also deepened and widened them, removing their capacity to support important habitat diversity. The resectioning (straightening and/or deepening of channels by excavating smooth, angled banks) of diverse river channels, often including removal of meanders to produce uni-form linear 'canals', may have served to optimise the rapid flow of water to the sea, but also robbed our rivers of their natural character and the many beneficial services that they could provide for society. This, allied with the building of flood banks and the disconnection of riparian floodplains, results in spates of muddy flows full of sediment after periods of heavy rainfall, rather than retaining water within wetland habitats and broader river corridors. The result was increased flood risks downstream as well as diminution of river flows during drier weather.

The extent of river management is dramatic. For example, a baseline survey of river habitats across England and Wales by the Environment Agency between 2006 and 2008 showed that almost 43% of the total length of rivers had been resectioned, largely associated with land drainage and flood defence activities.

Fig 12.10 For too long, we have over-managed rivers as simple drainage channels.

Furthermore, almost 8% of river channels had their banks reinforced with concrete, brick or other hard protection, generally associated with towns and cities or the protection of major roads and railways. In addition, loss of riparian habitat, including trees, has reduced not only the overall habitat structure of river corridors, but also important sources of food and organic matter. Shading of river channels has also declined, which exacerbates the impact of climatic changes, threatening raised water temperatures in future and a decreased capacity to carry dissolved oxygen.

12.7 Space invaders

These often substantial changes to the characteristics of British river and stillwater systems are further exacerbated by the invasion of problematic alien species. For example, the American signal crayfish (*Pacifastacus leniusculus*) is not only a significant predator of eggs and juvenile stages of fish, but also of the plants and the invertebrate communities necessary to sustain them. It has the ability to change ecosystem structure and characteristics through its dietary and burrowing habits.

Fig 12.11 The American signal crayfish is now widespread in British waters, causing significant impact across many local aquatic ecosystems.

Fig 12.12 Himalayan balsam, a common sight by many British rivers, is an aggressively invasive species that can displace native wildlife and destabilise banksides.

Invasive plant species on river banks, including Himalayan balsam (*Impatiens glandulifera*), Japanese knotweed (*Fallopia japonica*) and giant hogweed (*Heracleum mantegazzianum*), can change habitat structure and invertebrate and plant communities on river banks, and also exacerbate erosion as banks lie bare in winter when the vegetation dies back. Meanwhile, aquatic invasive plants, such as parrot's feather (*Myriophyllum aquaticum*) and floating pennywort (*Hydrocotyle ranunculoides*), can proliferate and shade out native vegetation and the invertebrate and other species that it supports.

As already mentioned, British waters have also been significantly affected by the activities of some foreign fish. This include the common carp (*Cyprinus carpio*), which can perturb ecosystems through their habit of grubbing in aquatic sediments which displaces rooted plants and also increases eutrophication, and also the topmouth gudgeon (*Pseudorasbora parva*), which breeds many times each year and may inhibit the recruitment of other native species. Invasive fish species may also bring with them exotic diseases that can infect native fish populations, which may lack evolved resistance.

Game fish species, too, both those introduced to the wider world from the British Isles as well as the many species brought to our shores, inevitably have consequences for the balance and functioning of the ecosystems into which they have been introduced, but within which they have not evolved. They may out-compete native species of fish and other organisms, prey upon them, spread parasites and diseases, perturb the habitat of ecosystems and may reduce the genetic diversity of native ecosystems where they breed faster than, or cross-breed with, native stocks.

12.8 Predators

Further pressures on native fish stocks include predators. Of course, predators are entirely natural elements of ecosystems, and the all-too-common clamour amongst some managers to blame predators for all the ills of their fisheries is often misplaced. A healthy ecosystem is one that is resilient, naturally evolved to absorb or adapt to changes in climatic, meteorological or extreme conditions, including the ravages of predator species, which serve to weed out the sick and excessive members of fish populations.

Where there is a perception that predation threatens the overall balance of ecosystems, the problems are very often more complex, with a few or multiple pressures creating a 'bottleneck' on the natural productivity of the system, which means that dwindling numbers of prey fish are more vulnerable to predation. For example, if poor habitat and water quality compromise the amount of recruitment of young fish to populations, or where habitat loss reduces refuge from both predation and spate flows, it is reasonable to expect that a boom in predator numbers

Fig 12.13 The proliferation of cormorants inland since the 1990s has undoubtedly increased pressures on the fish stocks of fresh waters.

may be the final metaphorical 'straw that breaks the camel's back', rather than the root cause of poor fish stocks. In these cases, the culling of predators may briefly satisfy those who clamour for a single diagnosis of the problem and a quick and visible fix, and there may be a little short-term relief, but this measure is unlikely to be a sustainable solution to fishery restoration.

There are, however, broader considerations about the impacts of predation. For example, large pike are highly efficient cannibals, controlling the numbers of juvenile pike with which they would otherwise compete for food. One of the consequences of pike culling, still commonly practised in trout fisheries, is a population explosion of juvenile pike, which, out of balance with the rest of the ecosystem and growing rapidly, may inflict increased damage on juvenile game and other fish stocks. Cormorants also attract a great deal of ire from some quarters of the angling community. Cormorant numbers in inland British waters have undoubtedly boomed since hunting bans have been in place, but perhaps the greatest impacts on the pervasion of cormorants inland has been the dual effect of depletion of coastal seas, their native feeding habitat, through excessive commercial fishing coincident with the construction of many 'inland oceans' (gravel pits and other large man-made water bodies). Here they find roosts and feeding habitat, particularly in waters artificially stocked to sustain commercial recreational fisheries. This encourages birds to fan out over wider landscapes, particularly so when some of these large freshwater bodies freeze over in extreme winter weather. Again, predator management may merely be a short-term 'sticking plaster on a gaping wound', sustainable solutions necessarily involving not only restoring the resilience of immediate freshwater fisheries, but also of the marine systems to which cormorants are better evolved.

12.9 Game fishes under pressure

There is no doubt that game fish species are under considerable stress. Indeed, where they have been introduced beyond the native ranges in which they have evolved as interdependent elements of ecosystems, they may themselves be major contributors to environmental stress.

Urbanisation, industrialisation, agricultural intensification and many other facets of our particular pathway of development combine to place a complex series of pressures upon aquatic ecosystems, including their fish fauna. When ecosystems slip out of balance, or where their resilience to cope with pressures is undermined, the integrity and functioning of those ecosystems is bound to diminish. When we lose characteristic species, including those which may have cultural and sporting significance and also act as indicators of the health of ecosystems, then we lose not only the ecosystems, but the many benefits they provide for us. If our salmon, trout, charr and grayling are under pressure, we can be assured that our own future wellbeing is becoming compromised by the progressive decline of the quality of the wider environment.

12.10 Under further pressure

This combination of issues arising from the use of urban and rural land, over and above direct impacts on the chemistry, hydrology and habitats of our aquatic systems, have placed huge pressures on our native wildlife, including such vulnerable groups of organisms as game fish species.

Pressures upon these fishes and their wider ecosystems are exacerbated still further by the modern fish-farming industry, to which we will turn our attentions in the following chapter.

CHAPTER 13

Down on the farm

Aquaculture has the potential to be a sustainable source of fish and other seafood, contributing to alleviating the significant and often overbearing pressures that we place on the world's oceans. However, like any intensive farming system driven by profit maximisation, often in ignorance of, or disregard for, the wider implications of farming activities, the sustainability of current stewardship of fish farms is highly questionable due to organic, nutrient, pharmaceutical and other pollutants escaping the farms, parasite and disease transmission, escaped fish, the nature of the feed itself, and a range of other problematic issues that we will touch upon in this chapter.

In the UK, the acceptability and market for cyprinid and other groups of freshwater fishes other than salmon and trout is low. This is somewhat at odds with the rest of Europe. Consequently, the culture of fish in Britain as well as Denmark, Norway, Canada and even Chile is dominated by species from the salmon family, supplying both the human food chain and stock for introduction to angling waters. British aquaculture includes the rearing and growth in fresh waters of brown trout and the rather hardier and faster-growing American rainbow trout, which has been taking place for many years, and more recently the farming of salmon in estuaries, lochs and sheltered coastal waters. (Grayling are not a farmed species and only limited success has thus far been achieved with Arctic charr.)

Furthermore, aquaculture is on the increase, estimated to provide us globally with more fish than are harvested from the world's oceans by 2030. Understanding and managing the associated risks and pressures is therefore a pressing priority.

13.1 Caged salmon farming

One of the major pressures on salmon and sea trout stocks across the British Isles is the culture of caged Atlantic salmon. Farming of salmon has grown rapidly, both in overall volume and in the size of individual farm businesses. For example, UN Food and Agriculture Organization figures note that commercial fisheries harvesting wild stocks produced 99% of salmon consumed worldwide in 1980, yet, by 2003, approximately 60% of salmon were produced in fish farms. In 2005,

Fig 13.1 Marine farmed salmon cages have become increasingly common on the Scottish west coast (image courtesy of Andrew Graham-Stewart).

salmon were farmed in 24 countries, with Norway, Chile, Scotland and Canada being responsible for 71% of global production, of which Atlantic salmon constituted the most economically important species representing 89% of salmon production for the human food chain.

The average size of salmon farming operations around the globe is increasing more dramatically than the number of farming operations, resulting in an industry increasingly dominated by a smaller number of large producers. For example, in Scotland in 1994, 19% of the farms produced over 1,000 tonnes of salmon yet, by 1999, this had risen to 59% whilst, over the same time period, the number of salmon farming operations had declined by 29%. Many adverse impacts flow from modern salmon-farming practices, including nutrient enrichment, habitat alteration and damage to wild fish populations. The use of wild fish in fish meal to feed farmed carnivorous fish, such as salmon, is a further significant problem.

Fig 13.2 Farmed salmon has become an increasingly common part of our diet since the 1980s.

The spread of sea lice (*Lepeophtheirus salmonis*) from caged salmon is a particular cause for concern. Sea lice, natural ocean parasites of Atlantic salmon and sea trout, graze the skin surface of the fish. However, where sea lice thrive on dense populations of caged fish close to estuaries and other important migratory routes, the dispersal of their planktonic larval stages can result in the infestation of juvenile wild Atlantic salmon and sea trout populations. When infested by more than a few lice, the parasite's grazing activities can reduce the resistance of salmon skin to infection. They also compromise the function of the gills, particularly their capacity to 'pump' salts into and out of the blood, and these overall stresses have

Figs 13.3 & 13.4 Intensive coastal salmon farms; as yet far from sustainable.

the potential to kill affected salmonids or, in the case of sea trout, to trigger an early return to freshwater. Sea lice infestations of wild stocks stemming from caged fish farms have been linked to dramatic collapses of returning wild stocks of salmon and sea trout. Other diseases are also a concern. This includes familiar diseases such as Infectious Salmon Anaemia (ISA) and also novel ones introduced with stock fish such as the fluke *Gyrodactylus salaris*, which has devastated wild and farmed Norwegian salmon; keeping it out of British waters is a pressing priority.

Escapee farmed salmon, some of which may be genetically engineered and all of which will lack the genetic diversity of wild fish, can also interbreed with wild stock. Farmed salmon, most of which originate from Norwegian stock, differ from

Fig 13.5 Sea trout severely infested with sea lice, with associated lesions.

Fig 13.6 Smolt farm at the top end of Loch Shin (image courtesy of Andrew Graham-Stewart).

wild salmon both morphologically and physically, which can affect behaviour, spawning success and competitive ability. Interbreeding between farmed and wild salmon has been found to be widespread, resulting in the dilution of salmon gene pools evolved since the end of the last ice age and lowering the overall resilience of hybrid stock.

Escapes do not only emanate from marine farms. Juvenile salmon are often grown in cages in freshwater lochs, and escapees easily mix with wild parr and smolts. Precocious male parr are particularly important in the fertilisation of salmon eggs, and therefore farmed escapee parr are quite possibly the most dangerous stage as far as interbreeding between wild and farmed fish is concerned.

It is also likely that future genetically engineered salmon, for example with additional genes with 'anti-freeze' properties or that may increase growth rate dramatically, may also further complicate this issue. Further concerns arise from the chemical wastes from caged salmon farms, including fish medications, elevated levels of heavy metals, nutrients such as phosphorus and nitrogen, and biological wastes with their associated nutrient loadings.

13.2 Trout farming

The farming of native brown trout had been a sporadic activity across the British Isles since the late nineteenth century, then largely based in enclosed pond systems with their associated water quality, disease, escapee and other problems. However, commercial-scale trout farming was introduced here in the 1950s by a Danish entrepreneur who had perfected the technique of rearing trout in elongated through-flow ponds, fed at the head by river or spring water and draining at the tail into watercourses.

Fig 13.7 Farmed rainbow trout, once considered a luxury and now commonplace due to aquaculture.

Freshwater cage farming is an alternative method, similar in many ways to caged salmon farming and involving placing net cages in deep freshwater lakes, with the Scottish lochs providing optimal locations for this method of farming in Britain. Since the 1950s, the trout-farming industry has grown significantly, achieving a total of some 360 trout farms across the UK, particularly in central and southern Scotland, the south of England and North Yorkshire. This distribution is related primarily to the requirement for an adequate clean and cool water supply in an accessible farm site.

Fig 13.8 Milt is stripped from a cock rainbow trout to fertilise eggs stripped from a hen for growing on in a hatchery.

Fig 13.9 Newly-hatched rainbow trout, alevins with yolk sacs attached, grow on in the hatchery.

A further transformation of the trout-farming industry in Britain, across Europe and more globally during this period has been the introduction, and now dominance, of rainbow trout. Though native to Pacific catchments of North America, these fish cope better than our native brown trout with the farming system, and generally grow faster. The British industry's trade organisation, the British Trout Association, recorded that around 16,000 tonnes of rainbow trout were produced in Britain in 2009, around 75% of which were marketed for human consumption and the remainder for sport fishery stocking and other purposes.

Issues associated with the pollution of water exiting fish farms have caused significant problems. Trout farm discharges are therefore now strictly controlled by licensed discharge consents. Other problems stemming from fish farms include the entrapment and consumption of fry of other fish species, and of planktonic and other aquatic diversity forming food important for river ecosystems. Antibiotics and other therapeutic chemicals are likely to be discharged in farm water effluent, often bound up in silt from fish ponds, which may also include spent food and waste products. Aquaculture also attracts piscivorous predators

Fig 13.10 Trout fry are fed to grow them on in the farm.

Fig 13.11 Trout are grown on in parallel ponds flushed with groundwater or river water.

that can then feed on the neighbouring river and lake systems. Diseases and parasites can be transported to neighbouring watercourses, especially by escaped stock fish, which, in turn, can occupy valuable wild fish habitat. These issues are not dissimilar to the problems caused by caged salmon farms.

Whilst trout farming should not automatically be considered as negative, especially as it provides valuable food as well as local income, it must be well managed if it is not to have potentially serious adverse consequences for stocks of native fishes of the salmon as well as other families, the wider aquatic ecosystem and all who depend upon or benefit from it.

13.3 Wild harvest

Pellets derived from marine resources are the primary feed used in commercial trout and salmon fish farming. They also have many other uses, particularly for recreational angling for carp, catfish and other target species. Pellets used as feed in salmon farming and their many other applications may be clean, nutritious and easy to handle at the point of use, but have a substantial associated 'footprint' of harm to the environment.

The bulk protein content (usually about 40% to promote rapid fish growth) of these pellets is derived from fishmeal. While some of this is derived from the offal and 'frames' of fish processed for human consumption, much is still manufactured from sand eels, capelin, blue whiting and many other global marine fish species, with the remainder accounted for by grain, soya, animal by-products, bulking agents and fish oils, all of which have their own associated environmental implications. Given the conversion ratio of energy and matter between trophic levels in food chains, it takes three to five tonnes of wild fish to produce one tonne of farmed salmon, trout or other predatory fish. When one considers the scale of contemporary trout and salmon farming, this represents a massive input of marine resources, and also extensive harm to coastal ecosystems given the far-from-sustainable practices of most of these fishing operations.

During much of our industrial development, characterised by relatively low human populations and inefficient capture methods, the sea must have appeared a boundless resource. However, as we observed in considering the impacts of commercial fisheries, the fishing power possible today through the use of petrochemical energy, radar, global positioning systems (GPSs), sonar fish detection and a host of other technological advances, means that the harvesting of marine resources, and the collateral damage to marine ecosystems, is carried out at an unprecedented and manifestly unsustainable level. We know that many marine fisheries are already being harvested well beyond their sustainable limits, threatening not merely the fish stocks themselves, but also species dependent upon them, and the wider ecosystems damaged or degraded by the impact of heavy trawls and other fishing equipment. For example, many marine bird species (such as Arctic terns, puffins, guillemots and razorbills) are undergoing significant declines in population, attributed to declining food resources. This, in turn, impacts their overall health, breeding success and resilience to other natural and

Fig 13.12 Most trout pellets comprise a large proportion of meal from sea fish.

man-made environmental pressures. Commercial overfishing, allied with by-catch mortality and other climate, habitat and wider environmental changes, plays a major role in declining seabird populations globally.

Whilst the five species of sand eels inhabiting the North Sea may be individually small, they swim in large shoals and are both abundant and important elements of coastal food webs across the north Atlantic. They are closely associated with sandy substrates, in which the fish burrow for much of their lives. Sand eels are important constituents of food webs, and their overall population decline has also been implicated in the deterioration of the vitality and productivity of other commercially important fish species such as cod. They are also an important food source for some marine mammals including the porpoise and the common seal. We should also not forget that sand eels, both adults and juvenile stages, are key constituents of the diet of the marine life stages of Atlantic salmon and sea trout, amongst a host of other marine life.

Sand-eel fisheries took off as other larger and more commercially valuable species of fish declined in numbers, largely as a consequence of overfishing. The sand-eel fishery at Shetland, for example, started in the early 1970s, with the highest landing of 52,000 tonnes recorded in 1982. Sand eels today support a huge international fishery in the North Sea, with annual quotas set as high as 400,000 tonnes in recent years.

ICES, the International Council for the Exploration of the Seas, established in 1902 as an intergovernmental organisation to '. . . promote and encourage research . . . for the study of the sea, particularly the living resources thereof . . .', recommended that local depletion of sand-eel aggregations by fisheries should be prevented, particularly in areas where predators congregate. They advised the closure of the sand-eel fisheries east of Scotland for 2000–2003 to allow populations of both fish and birds to recover. The UK Government has called for a moratorium on sand-eel fishing adjacent to seabird colonies, and a precautionary closure was established along the UK north-east coast in 2000. Notwithstanding such proactive measures, many voluntary environmental organisations, prominently including the Royal Society for the Protection of Birds (RSPB) and the Salmon & Trout Association in the UK, see the conservation of marine resources as insufficient, with reform of policy unacceptably slow and hampered by inadequate data. These organisations make the case for improved and ecologically driven marine protection, including such poorly controlled and inherently harmful activities as the overharvesting of sand eels.

The widespread use of fish-derived pellet feed in aquaculture therefore represents a largely unappreciated but undoubtedly major 'footprint' on marine ecosystems, and the many human interests supported by them, which must be factored into the assessment of the overall impacts of fish farming practices on ecosystems and the people who depend upon them.

13.4 The wisdom of ranching predatory fishes

Their preferred taste and traditional place in the British diet notwithstanding, many people are now beginning to question the wisdom of farming predatory fishes.

Occupying higher levels in the food chain, the conversion of feed into salmon and trout flesh is necessarily wasteful of energy and matter as well as producing excess waste, water pollution and other impacts. The feed conversion ratio (FCR: total feed fed/total weight gain) of farmed salmon improved considerably from the mid-1980s to the 1990s due to a transition in feed. This transition was from farm-made, semi-moist feeds (comprising minced fish, cereal flour and nutrient premixes) with an FCR of 4–6, a foodstuff gradually replaced by dry, commercially manufactured pelleted feeds with a higher protein and lower fat content leading to a much improved FCR of 1.6–1.8. However, although on-farm conversion improved threefold, this overlooks the substantial inputs of energy and materials expended in pellet manufacture, which may mean that the total conversion of raw resources into salmon body weight has changed little, if at all. Taking account of this wider input of raw resources, it is likely that approximately three tonnes of raw feed is required to produce a single tonne of fish flesh in ideal conditions.

We do not commonly rear predatory animals for the table, their potential ferocity and adaptation to kill prey being subsidiary factors to overall massive resource inefficiency. In addition, the climate change implications of human diets rich in meat, as compared to vegetarian options or a better omnivorous balance, is coming under closer scrutiny, not least by the British Government. Regardless of the now common place of farmed salmon and trout in the national diet, and the established vested interests of the aquaculture industry, the same principles apply to aquaculture.

Fig 13.13 Trout and salmon fillets hang ready for smoking.

Elsewhere in the world, the farming of more omnivorous species, including the common carp, tilapia and, increasingly commonly, Asian *Pangasius* catfish (known in the European market variously as 'pangas' or 'basa') may represent an inherently more resource-efficient form of aquaculture. Indeed, it is one that has persisted for centuries in China, elsewhere in east Asia and in the stewardship of fishponds by

European monks. Whilst aquaculture of these species is not without its own environmental, social and health challenges, their lower rank in food chains, together with the potential to farm local species, means that they are an inherently more sustainable option.

13.5 Seeking solutions

So, what can be done to make salmon and trout farming sustainable? The Salmon & Trout Association has a three-tiered policy – short, medium and long term – to overcome the adverse impacts of the aquaculture industry on wild fish and their environment.

Firstly, decision makers in governments and the fish-farming industry need to heed existing scientific evidence of this impact, for they have been in denial for too long. In particular, we know that farms sited close to river systems naturally populated by salmon and sea trout – in estuaries, lochs, enclosed coastal waters such as Scapa Flow on Orkney, and the immediate coastal zone – are having the greatest effect. These waters are generally sheltered, and provide perfect sites for aquaculture units. So, short-term measures must include the relocation of intensive farms away from the most sensitive sites through which wild migratory fish run. The only scientific problem left to be resolved is just how far away constitutes 'far enough'.

There is a view that the salmon-farming industry could relocate to massive units offshore, having researched new technology to allow the building of robust cages to withstand the weather. This is certainly a medium-term solution, but is it the final answer to achieving sustainability? After all, sea lice larvae have been known to drift seventy miles and, anyway, we still do not know sufficient about the migration routes of wild salmon and sea trout to be sure that they will be safeguarded. Even offshore sites could be slap in the middle of a traditional migration path.

The impact of escapees could certainly be minimised in the medium term by the industry moving within an agreed timescale to the mandatory use of triploid, sterile fish which, if they do escape, at least cannot interbreed with wild fish and dilute gene pools. Trout farming has moved this way in recent years and, interestingly, the Environment Agency in England and Wales has decreed that all brown trout stocked into rivers to support rod fisheries must be from triploid stock by 2015, with the intent of protecting what remains of our diverse wild brown trout populations. It could be argued that this is a little late in the day, as cultured trout have been stocked in rivers and lakes for at least 150 years and so any interbreeding has probably happened by now. However, the principle of protecting genetic heritage has at last been acted upon and, even if late in the day, this is an insurance for remaining wild stocks. If this is good enough for wild brown trout, then why not for wild salmon as well, since the principles at play are identical?

The long-term answer to minimising the impact of salmon aquaculture on wild trout and salmon populations is to operate farms with a biological barrier between wild and farmed fish. The way this is being researched at the moment very successfully, especially in Canada and China, is to grow the fish in closed containment units, consisting of enclosed tanks, either on land or immersed in the sea very much as netted cages, thus providing far more control over all aspects of the farming operation. The one downside is that energy is required to pump water through the tanks, although with less than a metre of head required, this is not excessive and renewable energy on exposed farm sites is a distinct possibility. The benefits are that escapes are minimised, no parasites can transfer between wild and farmed fish, disease is more easily treated than in open nets, and all waste products can be collected, processed and reused in most instances as agricultural fertilisers, rather than being allowed to contaminate the sea bed, as at present. And although capital outlay will be greater for these units, their longer lifespan and greater control will almost certainly pay dividends, not to mention taking away the need to treat fish for lice infestations, an extremely high cost for farmers conducting regular prophylactic doses to keep their stock healthy.

This also applies to the farming of trout and salmon smolts in fresh waters. The Norwegians, the world's largest producer of farmed salmon, already ban aquaculture units on waters containing wild fishes of the salmon family. So, freshwater cages must be removed from sensitive lakes and lochs, and land-based sites must be secure enough to minimise absolutely the possibility of escapes and the transfer of diseases and parasites.

Conservation NGOs are now promoting closed containment on a global level to combat the problems emanating from aquaculture. The industry could then present itself to the market place as a much more sustainable supplier of food, with obvious benefits for marketing to increasingly sophisticated and environmentally conscious retailers and consumers.

So, let us assume that fish farming can site itself away from impacting wild fish and the aquatic environment. Can we ever make artificial feeds for predatory species sustainable in terms of the natural resources required to produce them? It is possible, but there are problems. For instance, salmon and trout will grow on vegetable diets, but the resulting flesh will not contain the omega-3 oils that make these species such healthy eating; for that they need fish oils. This is hardly surprising, as nature has evolved them to be predators. The industry, to its credit, is researching new sources of protein to grow salmon and trout with their associated health benefits for human consumers but, for now, this still depends upon an unequal ratio of wild fishmeal to farm produce.

Aquaculture has much to offer us in terms of future food provision, potentially on a massive scale, but the industry, and the governments that support it globally, must face up to their responsibilities to ensure that farming methods, including feed input, are genuinely sustainable. If that is achievable, then the benefits to humanity, and the environment which supports us, will be most significant.

13.6 Searching for solutions: work in progress

Aquaculture plays a major and increasing role in supplying food and social benefits for humankind. It will therefore be essential to develop and assure means to carry it out without significant negative impacts on the environment and local communities. The Aquaculture Stewardship Council (ASC; www.ascworldwide.org) was established in 2009 with the mission 'To transform aquaculture towards environmental and social sustainability using efficient market mechanisms which create value across the chain'. Achieving this using market forces is central to the ASC strategy, operating by continuous improvement through compliance with standards at the farm level that can be audited through to consumer labelling at point of sale. This process is not only similar to the successful Forest Stewardship Council (FSC) and Marine Stewardship Council (MSC) accreditation schemes, established respectively in 1994 and 1998 to develop markets for sustainably sourced forest and fishery products, but was positively modelled on it. Like them, ASC was co-founded by the WWF (the Worldwide Fund for Nature), in the case of the ASC with IDH (the Dutch Sustainable Trade Initiative; www.idhsustainabletrade.com) as its other co-founder and with a growing network of partners including multinational companies.

It is early days, and it will be essential to continue to develop and test the standards of the ASC, which currently address aspects of compliance with local legislation,

Fig 13.14 The quest for sustainably farmed trout and salmon continues.

wider impacts on users, land, water and local fish stocks, feeding practices, fish welfare and contribution to local community development and poverty alleviation. It is essential to listen to and take seriously into account the concerns of NGOs and wider society as a means to build rigour and trust. Also, to report on compliance and continuing progress transparently and openly if the brand is to achieve credibility and market penetration, and to make a real contribution to sustainable aquaculture. By early 2012, standards were in development or implementation for farmed *Pangasius* and tilapia and also abalone and bivalves. However, if successful, the ASC will expand to cover other species potentially including trout and salmon, and hopefully address in practical terms the concerns raised in this chapter as the basis for creating trusted markets for more responsibly farmed products produced from them.

Meanwhile, in the USA, a new range of products produced from salmon farmed on a yeast-based feed rich in omega-3 fatty acids came on the market in September 2011. Use of the yeast-based feed can cut down by 75% the amount of wild-caught fish pellets fed to farmed salmon, requiring just 1 kilogram of fish-based feed to produce 1 kilogram of salmon compared to around 4 kilograms for traditional fish-based feed. Yeast-fed salmon currently have a price premium, perhaps reflecting how little we currently value impacts on the natural environment of fish-feed production. However, more breakthroughs of this kind will be necessary to retain the health benefits of farmed trout and salmon whilst offsetting associated negative impacts.

In a further development, the Research Council of Norway looked at the future of marine fishing in 2003, an outcome of which was a new programme to make Norway '. . . the world's leading aquaculture nation'. This includes developing vaccines and medicines to deal with sea lice, which clearly would reduce risks of transmission to, as well as from, wild fish. Progress of this kind is welcome, provided full sustainability is both the longer-term goal and a clear commitment for continuous improvement.

Meanwhile, WWF have a vision of 'integrated multi-trophic aquaculture' including, for example, farming fish in proximity to algae, which may take up nutrients excreted by the fish. The algae may then be consumed by farmed shellfish, or else harvested for biofuel production. This integrated approach has promise, though it remains as yet far from practical implementation. The need to provide food with reduced impacts on the environment and people will clearly be a priority; aquaculture will undoubtedly have a role to play in future food security, and so these and other developments are welcome.

All of this must still be considered as 'work in progress'. The underlying issue with all of aquaculture has to be that, however large and successful the global industry becomes, its impact on the natural world and its dependent species must be minimal for it to be considered genuinely sustainable.

Chapter 14

Salmonids under pressure

The systemic changes to the landscape and aquatic ecosystems of the British Isles, and pressures on the coastal and high seas used by migratory fishes, have been occurring over centuries. However, the pace of change since the age of industrialisation has accelerated dramatically, and is continuing so to do. As we know from the extinction of British wildlife at the hand of humanity, such as the large tortoiseshell butterfly from our woodland and the burbot from our rivers, there are natural limits to the tolerance of all species. If this can be the case for the now extinct passenger pigeon, once the most common bird in North America that was formerly so abundant that its migrating flocks took literally hours to pass, it can certainly be true for any other wildlife.

Fig 14.1 Obstructions to the free passage of salmon, trout and other migratory fish up and downstream pose significant threats to local fish populations.

As denizens of high-quality water environments, the British game fishes are highly susceptible. This is particularly so for migratory species, such as Atlantic salmon and sea trout, which use networks of connected habitats across which they are vulnerable to threats from chemical pollution, habitat disturbance, obstructions, overexploitation, imbalances of predators and parasites, competition from introduced fish, loss of genetic integrity and many other pressures besides.

14.1 The fortunes of Atlantic salmon

Atlantic salmon are still struggling in most English and Welsh rivers. The Tyne has made a great comeback due to cleaning up the river from its industrial heritage, and 4,638 salmon were caught from it by rod-and-line anglers in 2010, 42% of which were multi-sea-winter fish. Whilst the Kielder Hatchery served as a useful catalyst for this recovery, one of the major factors implicated in the return of salmon was the significant reduction in the north-east coast drift-net fishery. On the basis of fish counter data, the mean number of upstream migrants on the River Tyne in the eight years since the buy-out (2003–2010) has been 36,936 fish (salmon and sea trout combined), compared with an average of 21,152 fish in the five years prior to the buy-out, representing an increase of 75%. The count in 2010 (45,602 fish) was over twice the count in 2009 and was among the highest in the recent time series.

The Tees would have a decent run of Atlantic salmon if it wasn't for the barrage that was completed in 1995 just upriver of Blue House Point. Fixing the problems caused by this obstruction is, in fact, a classic case of human prevarication. The report of a two-year study by Cefas (the UK Government's Centre for Environment, Fisheries and Aquaculture Science) confirms anglers' fears that the barrage represents a near-impassable barrier to migratory fish as, of 72 fish electronically tagged in the estuary, not a single one successfully negotiated the barrage to migrate up river to spawn. Up to 76% were killed by seals, which had learned of the rich pickings of waiting fish at the foot of the small fish pass, and the remainder either disappeared back out to sea or were untraceable. The report also recorded low dissolved oxygen and high ammonia levels in the estuary, often at levels deemed unsafe for aquatic organisms, which therefore further reduce the fishes' ability to evade predators and complete the assault course presented by the barrage. Meanwhile, the salmon stacking up below the barrage are vulnerable to the frequent activities of poachers. Notwithstanding all the legal protection that Atlantic salmon theoretically enjoy, it is hard to understand why the questionable economic interests that saw the design and construction of the barrage are still getting in the way of adequate conservation of this iconic fish and the wider ecosystem.

Other English rivers still enjoy reasonable runs of salmon. These include the Eden (1,407 caught in 2010 of which 32% were multi-sea-winter fish), the Lune

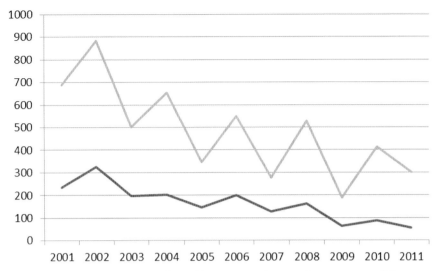

Fig 14.2 Monitoring data suggest that wild migratory salmon and sea trout are still in long-term decline.

(1,069 including 20% multi-sea-winter fish), Ribble (1,145 of which 29% were multi-sea-winter fish) and the Derwent (1,142 of which 20% were multi-sea-winter salmon). Other rivers have smaller but significant runs, including several in south-west England, for example the Tamar, which saw 365 salmon caught (16% multi-sea-winter) in 2010, but these are still well short of historical numbers. The southern English chalk streams, especially the Hampshire Avon (just 35 caught in 2010), are shadows of their former selves. However, a few rivers – including the formerly polluted Mersey and Trent – are seeing a resurgence in Atlantic salmon runs, though these are from a dreadfully low ebb; we will need to see the trend continuing well into the future before we can get too excited about the extent and sustainability of recovery.

The Teifi (841 salmon caught) and Tywi (827) were the highest-performing Welsh rivers in 2010, although there are several others with viable runs. However, Wales is better known for its sea trout fisheries, with the Tywi, Teifi and Dyfi all in the top six English and Welsh rivers in 2008 according to the Environment Agency's fishery report.

In 2009, 30,204 sea trout were caught in English and Welsh rivers, of which the Tywi accounted for 2,845 and the Teifi 2,076, as opposed to the Tyne's 1,406 and the Wear's 1,271.

In Scotland, the picture is decidedly inconsistent. The spring run has rallied in recent years, although 2009 was particularly poor. However, it is unclear if this is part of a cyclical phenomenon. The grilse run has been poor in the latter part of the first decade of the twenty-first century, and many returning fish have been small and in poor condition. However, 2010 saw a bumper catch of salmon and grilse, with just over 110,000 being caught by the rods. The Tweed had a record year, with more than 24,000 fish being landed on that one river alone. The inconsistency continued into 2011, though, where grilse runs were poor again but multi-sea-winter fish increased in numbers, especially in terms of summer salmon.

Poor marine survival appears to be the major factor, but what is causing it? Are the fish dying because of changes in currents and temperatures, interference with the North Atlantic Oscillation, or other such meteorologically driven factors? Or is the wider marine ecosystem altering to the extent that prey species are collapsing and so the salmon are limited in food supply? Or, as many now suspect, is the cocktail of toxins, built up sublethally during the freshwater phase, killing migrating smolts as they hit the extra stress of entering the marine environment? We do not know enough about the Atlantic salmon's marine phase to understand all the complexities, which is why initiatives such as the EU-funded SALSEA-Merge project (described in Chapter 3, Brown trout or sea trout) are so important in collecting data from post-smolts and adults at sea. However, we need to know far more about what happens to smolts in the coastal zone as well.

14.2 Sea trout in decline

Sea trout are generally in decline everywhere, and we do not know why. We know that the Finnish sea trout population was brought to its (metaphorical) knees as a result of the by-catch of the coastal net fishery. Is this happening around the UK as well, particularly with the increase in gill netting for bass, mullet and other such inshore species? We also know that tiny upland streams are crucial to sea trout spawning, so has agricultural 'improvement' impacted juvenile habitat and, therefore, smolt production?

In Scotland, aquaculture on the west coast has certainly hit sea trout through lice transfer, but east coast sea trout stocks have been hit as well. Again, we need far more data. Three significant sea trout projects were under way in 2012 to explore this, including the Celtic Sea Trout Project, a collaboration between Wales and Ireland that should provide vital information better to inform future management decisions.

14.3 Trends in brown trout

Wild brown-trout stocks have undoubtedly suffered from habitat loss and, to a lesser extent, angling pressure, although 'catch and release' has generally lightened angling impact over recent years. Abstraction and agricultural pressures are probably the greatest causes of decline, along with physical changes to habitat.

Fig 14.3 Brown trout are hanging on in our rivers, yet remain subject to a wide range of pressures.

Brown trout often migrate to headwaters and tributaries to spawn and, like sea trout, they are vulnerable to drainage, over-grazing and abstraction, which combine to degrade or destroy necessary habitat. The stocking of hatchery trout for the past hundred years or so may also have caused significant dilution of gene banks. However, this is by no means proven and, many suspect, is often used to cover up more direct causes of wild fish population collapse as a result of activities sanctioned, subsidised or licensed by governments and their regulatory bodies.

14.4 The lady of the stream

Declining water quality in the UK has threatened the vitality of grayling populations throughout the past century. However, better treatment of pollution and the spread of grayling to new river systems have both done much to arrest this former decline.

Indeed, grayling have, arguably, made a comeback, principally because they have become a more appreciated fish by fishery managers and anglers (with particular praise due to the Grayling Society for promoting this). Where once trout fishery managers and anglers would openly discriminate against grayling, netting or electro-fishing out huge numbers, these graceful fish are now seen as a positive benefit and stocks are as healthy now, probably, than at any time in the past fifty years. Assessments undertaken by the UK's Joint Nature Conservation Committee (JNCC), published in 2007 to determine the status of the species under the EU Habitats Directive, expressed no concern about threats to this species.

14.5 Down in the cold depths

Arctic charr are struggling, and can only continue to do so, one suspects, in the face of climate change. Already severely depleted in Windermere, their Scottish haunts are also under pressure, as are Welsh populations many of which are also threatened by sewage and agricultural eutrophication of deep lakes. However, the lack of an extensive fishery for charr makes actual trends difficult to monitor.

The same observation also applies to the whitefishes, the vendace and the European whitefish (pollan, powan, schelly and gwyniad), climate change and nutrient enrichment combining with the impacts of introduced coarse fish species to pose significant threats to these nationally-scarce fishes.

14.6 A changing climate for salmonids

By far the largest threat to the UK's game fishes is climate change, and how the world, the UK and also fishery managers react to both mitigate and adapt to it. Salmon in southern England will, almost certainly, struggle to retain viable populations. Resident trout will have to be cosseted with improved bankside shading and in-river cover, as well as a serious rethink about abstracting water directly from rivers and groundwater supplies at current excessive rates.

At sea, we can only hope that changes in the timing of plankton growth and its species composition, as well as impacts on the wider food web and ocean currents, will not make the marine phase of the salmon and sea trout life-cycle any more stressful than at present. This could tip the balance against their overall survival. However, some populations will almost certainly survive, as they have done over millions of years past, and their ability to stray and recolonise rivers as conditions improve will surely protect them into the future – always provided, of course, that the many wider human impacts on the planet can be kept within sustainable limits.

Fig 14.4 Climatic instability may pose overriding problems for our fish and other wildlife.

Part 3

Game fishes for the future

CHAPTER 15

Sea change

Increasingly throughout the latter decades of the twentieth century, various high-profile events have emerged to shock the industrialised world from the utilitarian pathway upon which it has been embarked since at least the late-eighteenth century. Many of these surprises were environmental in nature, generally resulting in actual or potential humanitarian catastrophe.

Cumulatively, these unanticipated events served to loosen our blinkers, opening the door to new perspectives on the linkages between our economic activities, the supporting environment and its natural resources, and their inexorable consequences for human wellbeing. Environmental awareness flourished throughout the 1960s and into the 1970s, though the roots of this revolution were laid down long before that.

15.1 Water as a medium for environmental consciousness

Water is crucial to humanity, whether to support our basic biophysical needs for drinking, cooking and washing, or for irrigation of crops, supplies for our industries, mill-based and hydroelectric energy, transport of goods and defence from enemies, or the maintenance of ecosystems and good air and wider environmental quality. Water has also shaped the landscapes we inhabit and the uses to which we have put them, including often lending its properties to the names and character of human habitations, and the technologies that we have exploited. This even includes the innovation of the 'spinning jenny', the machine that heralded the dawn of the Industrial Revolution, which became known as the 'spinning mule' when coupled with a 'water frame' to harness the power of flowing water. Water has both promoted and limited the pathway of human development throughout our history.

It is, then, hardly surprising that water, including its conspicuous aquatic life, has also featured prominently in the evolution of consciousness about our intimate dependence upon the ecosystems of this planet. Flood, drought and famine, transmission of waterborne diseases and parasites, and the loss of or threats to charismatic flora and fauna have all mobilised public opinion and, sometimes, political response. Our increasing collective environmental literacy has shaped, and continues to influence, contemporary consciousness, such that no longer can those of us in the better-educated developed world claim ignorance about the unintended consequences of our historic pathway of industrial and intensive agricultural development. Nor, with pervasive global real-time media, can we turn a blind eye to the worldwide consequences of our cavalier disregard for the capacities of nature to sustain us, and the impacts already being felt by many sectors of society consequent from the unravelling of essential supportive ecosystems.

15.2 Dawning realities and the seeds of cultural revolution

Measures eventually leading to establishment of what is now the UK's Royal Society for the Protection of Birds (RSPB) constituted a significant, high-profile early step in society's transition to greater environmental responsibility, and was closely connected with safeguarding charismatic yet vulnerable aquatic life. The RSPB has its roots in a gathering of fashionable ladies who realised that their lifestyle demands were contributing to the loss of a water bird. The great crested grebe is today widespread on the larger slow-moving and still water bodies of the British Isles but, by 1860, it had been hunted nearly to extinction – down to just

fifty breeding pairs – for the trade in two fashion accessories. 'Grebe fur', the bird's skin and soft under-pelt, was popular as a fur substitute in ladies' fashions, and the head frill feathers of the adult bird's breeding plumage were becoming fashionable in the millinery trade. Mobilisation of public and political opinion to counter this trade was the foundation of the Society for the Protection of Birds, formed between 1889 and 1891 and gaining a Royal Charter to become the RSPB in 1904. Today, the RSPB constitutes one of the UK's largest conservation charities and is an influential environmental lobbying organisation within excess of one million members.

Fig 15.1 Once seriously endangered, the great-crested grebe has made a dramatic recovery as a result of changing environmental consciousness and concern.

At the same time, similar sentiments were taking root and crystallising across the Atlantic. The year 1905 saw the incorporation of the National Audubon Society in the USA, a charity named in honour of the French-American ornithologist, John James Audubon, reflecting not only his fame as a naturalist, hunter and painter, but also his prophetic warnings of the threats of over-hunting and loss of habitat for many once-common American birds.

Dawning awareness of the connections between human activities and environmental interests, that had formerly implicitly been assumed to be independent, can be traced in early evocative writing warning of environmental degradation or catastrophe. One of the earliest examples of this literature is Harry Plunket Greene's 1924 *Where the Bright Waters Meet*, alert to implications of the mismanagement of rivers and the landscapes in which they flow for the vitality of trout populations in his much loved Bourne Rivulet fishery in Hampshire. Across the Atlantic, Aldo Leopold's call for a new 'land ethic' in his 1949 book *A Sand County Almanac* made a compelling case for the extension of ethical standards from human kin to the ecosystems we share and which, in turn, support our continuing needs.

As the century matured, compelling evidence of the adverse effects upon wildlife of persistent pesticides, collated in Rachel Carson's seminal 1962 book *Silent Spring*, presented a shocking picture of the consequences of pesticide bio-accumulation, affecting common farmland birds in Britain and contributing to the mortality of arctic sea birds remote from where those pesticides were being applied. This alarming awareness was backed up by the first sight of our small, seemingly vulnerable home planet viewed from space in photographs beamed back from the Apollo 8 lunar mission in 1968. At the same time, we witnessed increasingly frequent, high-profile environmental disasters throughout the 1960s and 1970s, including the break-up and massive oil spill from the *Torrey Canyon* oil tanker, the world's first 'super-tanker' accident, off the south-west coast of England, and heavy metal contamination of marine ecosystems and the human food chain resulting from long-term discharges from a chemical plant into Minamata Bay in Japan.

These and other stimuli lay behind a cultural sea change with respect to humanity's conception of our relationship with the environment. For example, instigation of the UN's 1972 Stockholm Conference on the Environment elevated public and political awareness of the global nature of the environmental problems unwittingly caused by humanity, their longer-term implications for both eco-system integrity and human development, and the need for concerted action to find solutions.

15.3 Realising the connections

The key trigger for all of these initiatives was the drawing of linkages from water and other ecosystems to the fortunes of increasingly broad sectors of society.

The intimate and inseparable interdependence between humanity and the planetary ecosystems with which we evolved is, in terms of our cultural evolution, a relatively recent phenomenon, albeit that some aspects of it are embedded in traditional practices of resource use across much of the less-developed world. We are progressively learning that technological interventions that maximise only a few short-term benefits accruing from the way we use ecosystems, such as through unregulated aquaculture or intensive farming practices ignorant of the long-term fertility, biodiversity and hydrology of landscapes, have inevitable longer-term and cross-disciplinary consequences. The pathway of the developed world has, unfortunately, been largely built upon such technologies, which support a burgeoning human population beyond the natural carrying capacity of the landscapes we inhabit. We are faced with a pressing challenge to embark on a new journey of discovery of new technologies, and the rediscovery of old ones to support the needs of people indefinitely into the future. We have enough knowledge to appreciate, for example, that we face serious problems and that, by putting our waters, soils and their ecosystems under pressure, we are decreasing the capacity for all in society to lead a decent quality of life.

Fig 15.2 Recognition of our inseparability from water and the wider natural world is driving a new world view.

As we have already seen, water resources and their associated characteristic fish fauna and other ecosystem attributes, are amongst the most vulnerable, yet most important to sustaining continued human wellbeing. It is not without good reasons that the Millennium Development Goals – the UN-brokered promise of the already-developed world at the turn of the current millennium to those countries still held back by grinding poverty – revolve around water resources and their associated biodiversity and capacity to support a host of human needs. Nor indeed that South Africa's far-sighted National Water Act of 1998 explicitly posits the reallocation of water in this arid land as a means for reversing centuries of racial inequity and division.

With the salmonid fishes of the world, amongst other iconic and valuable fish fauna, also swims the potential for thriving aquatic resources to continue to sustain many attributes of human wellbeing into our longer-term future.

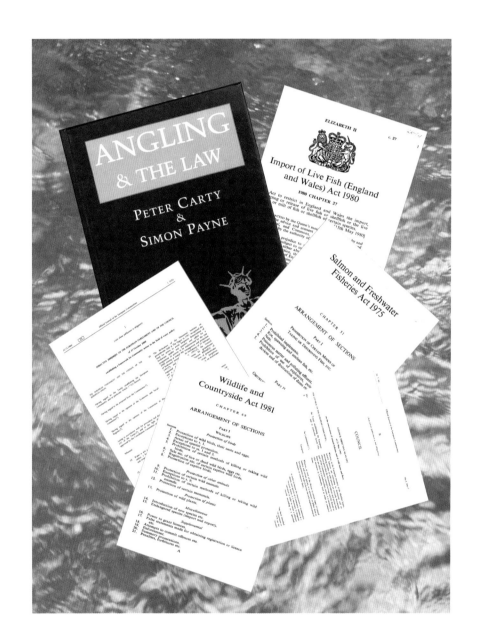

CHAPTER 16

Changing rules

The importance of rivers and their ecosystems, of which natural populations of fish including those of the salmon family are prominent and culturally important constituents, has been recognised increasingly over time. However, over the same period, societal pressures and habits have continued to put mounting pressure on the habitats essential for their sustenance.

It is in legislation, including associated subsidy and 'best practice' initiatives, that society cements changing norms. The instigation of these changing rules is rarely, if indeed ever, a top-down process of political leadership. More often, it is a slowly formed consequence of changing scientific, cultural, business or other awareness ultimately requiring the institution of new rules to guide the broader conduct of society. The democratic process is slow, and also subject to the views and influence of many voices, including those of vested (often economically and/ or politically influential) interests in established practices, for which change may be less welcome no matter how relevant it may be to longer-term societal equity, security and wellbeing.

Nevertheless, in the historical trace of regulatory reform, we do see hard evidence of the transitions that society has made. Many of them are, unsurprisingly, steered by the consequences for our collective wellbeing stemming from their impacts upon aquatic ecosystems and other fundamental natural resources.

16.1 Changing the flow

One should not be blinded by optimism about the difficulty and pace of cultural change, nor conversely stunned into inertia by the scale of the challenge. Modern society rests upon two-and-a-half centuries of entrenched habits and vested interests with respect to resource use patterns, economic incentives and business governance models that originally conceived ecosystems as boundless resources to be exploited competitively for short-term gain. We are, as we have already discussed, aware today that the world's resources are far from limitless, and that the escalating demands of a spiralling human population require radically different assumptions to those that shaped inherited and 'business is usual' trade, property law, financial systems, short-term electoral cycles and seemingly every other facet of the modern world. Substantial and rapid change is necessary, given the enormity and immediacy of the threat from climate change, declining ecosystems, global poverty, and the danger of conflict arising from competition for dwindling resources.

Yet the threats to our game fish, and also to ourselves via the environmental media upon which we all depend, require us to intervene not merely in some local habitat modification or anti-poaching activities (important though those measures may be) but in the progressive rewiring of whole complex sociopolitical mechanisms by which we govern our relationship with the ecosystems that ultimately support us. This challenge – putting sustainable development into practical action and at the heart of strategic governance – is the defining one of our age. The long-term viability of our game fishes, along with our own potential to live fulfilled lives, depends upon it.

A change in awareness and mindset concerning environmental impacts, and also their consequences for our own long-term wellbeing, may challenge established and vested interests. However, it also creates opportunities for new markets and different ways of viewing and governing our activities.

16.2 Changing the ground rules

Legislation at many different scales – international, national and local – has progressively tightened in the wake of the shifting perceptions and concerns held by society. This includes many instruments to protect vulnerable species, including several of the game fishes.

The legacy of water-quality legislation in the UK has been largely founded on the need to protect the health of both the public and of fish populations. From the late 1970s, with further revision in the late 1990s, management of river quality was explicitly linked to the needs of populations of different groups of fish and other aquatic life. The current 'River Ecosystem' classification system used to establish chemical quality objectives for all river reaches downstream of headwaters across England and Wales is based on the requirements of salmonid (salmon family) and cyprinid (carp family) fish populations.

Specific fishery-related legislation protecting Scottish salmon fishing is believed to have been in place before the eleventh century, but the first recorded legislation for salmon was passed by the Scottish Parliament in 1318. A great deal of legislation has been enacted subsequently across the UK, significantly including the Salmon and Freshwater Fisheries Act 1975, the Salmon and Freshwater Fisheries (Protection) (Scotland) Act 1951 and the Salmon Act 1986.

Scotland arguably led the way in modern salmon management when, in 1962, the fledgling drift net fishery in home waters was immediately recognised as

Fig 16.1 The needs of fish populations form the backbone of a great deal of pollution control legislation.

being potentially damaging to the overall salmon stock, and was banned without compensation being paid to commercial fishermen. As will be seen below, it took the rest of the UK and Ireland much longer to react to the danger, with significant damage being inflicted during the intervening years on many home salmonid rivers.

Following a thorough review of Scottish Fisheries Legislation between 2003–2006, including a forum system that gave angling and fisheries organisations and individuals unique input to the process, the Salmon, Fisheries and Aquaculture Act of 2007 consolidated and updated Scottish fisheries legislation. This proved how effective NGO involvement could be in bringing about legislative change. The value of involving NGOs and independent specialists, in addition to the usual government agencies, had become evident a little earlier south of the border. In England and Wales, a review of fisheries legislation took place between 1997 and 2000, when a committee representing every major sector of freshwater fisheries and angling, chaired by Professor Lynda Warren, looked into all aspects of law and management. NGOs played a leading role in the review, with thirteen angling and fisheries organisations coming together under the chairmanship of Lord Moran, then chairman of the Salmon & Trout Association, to form the Moran Committee, which gave evidence at every stage of the review. The Warren Committee made 196 recommendations in its report in 2000, and the government accepted the vast majority of these in their 2001 response. The Moran Committee was one of only two organisations individually recognised in the government response (the other being the Environment Agency), signalling significant NGO influence in the review's outcome. Many recommendations have been enacted through secondary legislation, but the Marine Act of 2009 is, arguably, the final piece of the jigsaw, with 15 clauses slotted into the middle covering freshwater fisheries issues.

However, probably the highest profile 'win' for NGOs from the review was the recommendation that mixed-stock salmon fisheries should be phased out in England and Wales; something for which the Salmon & Trout Association and partner NGOs had been lobbying over many years. These fisheries exploit fish from more than one river system, so rendering impossible the management of

Fig 16.2 Pressures from angling bodies and interests can be significant in advancing the health of waters and their many wider societal benefits.

stocks from individual rivers. The UK Government agreed with the recommendation and, incredibly, put £1.25 million of funding on the table to help buy-out licenses in the most damaging fishery, the north-east English coast drift nets, and so helped to speed up its eventual closure. With donations from clubs, organisations and individuals in Scotland and England, and negotiations undertaken by the UK-based section of the North Atlantic Salmon Fund, 52 of the 68 licenses were bought out in 2003, and exploitation of salmon dropped markedly in the fishery.

However, perhaps the NGO win that was most significant for salmon con-servation was the closure of the Republic of Ireland's drift-net fishery at the end of the 2006 season. The legal process leading up to this began years earlier with a complaint by the Wessex Salmon and Rivers Trust that mixed-stock salmon exploitation off the Irish coast was intercepting fish bound for the River Avon in Hampshire, England, which had been designated a Special Area of Conservation (SAC) under the EU Habitats Directive for its Atlantic salmon population. The cause was taken up by the Stop Now! campaign in Ireland, set up in 2004 by the Federation of Irish Salmon and Sea Trout Anglers (FISSTA) to establish a united angling and stakeholder group to lobby politicians. Stop Now! argued that Irish rivers designated as SACs were also being impacted by the fishery. The campaign received widespread support from all concerned UK NGOs, especially during the annual North Atlantic Salmon Conservation Organisation (NASCO – described later in this chapter) meetings between 1999 and 2005, where the Irish Republic, as members of the EU Delegation, came under increasing pressure to act. Once again, strong NGO pressure bore conservation results, with benefits felt well beyond merely the angling community. The eventual decision by the Irish Government to adopt the Independent Salmon Group Report to end mixed stock fisheries immediately was widely welcomed, together with the establishment of a com-pensation fund of €30 million to pay commercial driftnet fishermen over five times their annual estimated salmon income in order to ensure that Ireland would no longer be in breach of the EU Habitats Directive on this issue by 2007.

The body of environmental legislation affecting the game fishes is broad and diverse, and is also under constant revision. Recent momentum includes the consultation by Defra (the UK Government's Department for Environment, Food and Rural Affairs) in England, closing in April 2009, on modernisation of legislation to address the passage of migratory fish through obstructions and past abstraction and discharge points. However, because the hydropower industry was beginning to apply for permission to install renewable energy schemes on rivers supporting migratory trout and salmon, Defra delayed the legislation until 2012 over concerns that commercial interests would be impacted by the requirement to provide fish passes at every hydropower weir. Yet again, natural resources have to take second place to industry, even though the potential impact of hydroelectric generation threatens to jeopardise firm commitments to achieve good ecological status in rivers under the EU Water Framework Directive, and there are heavy fines under European infraction if these binding objectives are not met. Furthermore, clear signals about the free passage of fish would provide an unambiguous strategic direction around which hydropower interests could innovate. If ever there was an example of a lack of co-ordination between the different sectors operating within the aquatic world, the hydropower issue is surely up there with the worst of them.

The bulk of environmental legislation impinging on the UK now emanates from Europe. This goes back to the European Community (EC, as the EU was then known) Freshwater Fish Directive of 1978, which, though now repealed in the light of subsequent legislation, put in place mandatory physical and chemical quality standards to protect designated salmonid and cyprinid fisheries across EU Member States. The Habitats Directive, based on the 1979 Bern Convention concerning the conservation of European wildlife and natural habitats, was initiated by the EU in 1992. The Atlantic Salmon is scheduled under Annex II of the EU Habitats Directive, though in fresh waters only, and the grayling is listed as a species of 'Community interest' and as such its exploitation may be subject to fisheries management. However, brown trout and Arctic charr are covered only in the context of certain habitat types under the Directive, although they were given some protection by being made a focal species under the UK's Biodiversity Action Plan in 2010.

The EU Water Framework Directive (WFD) came on stream in 2000 and is the most comprehensive piece of European legislation yet enacted, covering all aspects of the water environment and, for the first time, placing 'good ecological status' as the intended outcome. Importantly, the WFD looks at the overall health of aquatic systems, including both surface waters and groundwater, rather than addressing specific elements in isolation. Atlantic salmon, trout, Arctic charr and grayling, as well as the whitefishes, are important indicators of the good ecological status of certain types of aquatic ecosystems, and indeed the connections between different habitat types. Their vitality, along with that of the wider aquatic ecosystems

Fig 16.3 The Water Framework Directive addresses many aspects of the health of aquatic ecosystems upon which fish, the food chains that support them, and human wellbeing depend.

of which they are a part, depends upon effective management of whole drainage basins and the many human activities that occur within them. This requires a new paradigm of management, with ecosystem health as a clear end-point. This is consistent with the commitments of the UK, amongst virtually all nations across the world, to a pathway of sustainable development. Indeed, amongst the 'headline' and supporting sustainable development indicators adopted by the UK Government are a number that pertain to biological diversity, access to green spaces, and other dimensions of environmental health and access. Some, such as those pertaining to water quality and population trends in wetland birds, relate directly to the overall vitality of water resources, whilst (as discussed later in this chapter) one relates directly to the health of Atlantic salmon populations.

The actual and potential problems associated with the introduction of alien species of fish and other organisms beyond their natural range, and the ecosystems with which they have coevolved, have already been described in Part 2 of this book. The import of fish into the UK is controlled under legal instruments including the Import of Live Fish (England and Wales) Act 1980 (ILFA) and the Import of Live Fish (Scotland) Act 1978, both of which take a precautionary approach of prohibiting import of fish species assessed as capable of forming self-sustaining populations in British waters.

However, as noted already, the political process requires established interests to be taken into account and given a voice in final decisions, which means that the pace of legislative change is inherently slow. Therefore, no matter how high the principles upon which overall intentions may be founded, established interests will tend to conspire to constrain the scale and pace of legal implementation and eventual cultural change.

The question is, do we still have this luxury of time to make the substantial and fundamental changes necessary to reverse humanity's more destructive habits?

16.3 The role of international conventions

Much conservation work has to happen across borders. This is particularly so for migratory fishes, as well as those species traded internationally.

Recognition of the potential threat of global trade for wildlife, worth billions of dollars and a valuable source of foreign revenue annually, particularly for developing countries, led to international co-operation culminating in the signing in 1973 of the intergovernmental Convention on International Trade in Endangered Species of Wild Fauna and Flora (CITES). As of 2009, there were 175 signatory nations to CITES, safeguarding trade in a variety of animals, plants and derivative products and constituting the only widely implemented international instrument that deals with sustainable international trade in wild species. Although CITES lists only three species from the salmon family (the Mexican golden trout and two scarce whitefishes), the convention demonstrates transitions in international concern and co-operation, augmenting national and supranational (including European) legislation to protect species of conservation concern.

As already described, major advances in fish location and other fishing technologies have had dramatic implications for the scale of capture of marine fish, accelerating the inherently unsustainable demise of stocks. Migratory salmonid fishes, particularly the Atlantic salmon, have been directly affected. Up until the 1960s, exploitation of salmon in the North Atlantic was within territorial waters and therefore occurred largely at a national level and could be better related to the health of stocks in natal rivers. However, development of coastal shelf fisheries at West Greenland and in the Northern Norwegian Sea, netting maturing salmon spawned in rivers from many different countries, meant that sensible management could only be achieved through international co-operation. This initiated the establishment of the Convention for the Conservation of Salmon in the North Atlantic Ocean. The convention entered force in October 1983, also creating the North Atlantic Salmon Conservation Organisation (NASCO) as the inter-governmental organisation charged with the objectives of conserving, restoring, enhancing and rationally managing wild Atlantic salmon.

Throughout the intervening years, NASCO has sought to remain current with scientific understanding and conservation priorities, its 2005 *Strategic Review* prioritising the restoration of abundant Atlantic salmon stocks throughout the species' range with the aim of providing the greatest possible benefits to society and individuals. The emphasis on wider societal benefits, rather

Fig 16.4 International conventions are all part of agreements to help Atlantic salmon parr continue to prosper.

than focusing on preserving stocks for more narrowly defined conservation and exploitation goals, captures some of the transition in conservation thinking over recent years. Part of this transition was an agreement to work more closely with NGOs to achieve longer-term goals, and NASCO now has more input from NGOs than any other international fisheries forum allows, with the Chairman of the NGO Group sitting at the 'top table' alongside the Heads of Delegation at annual meetings

16.4 A focus for co-ordinated government action

National governments, including that of the UK, also undertake a number of initiatives which, although lacking statutory force, are intended to co-ordinate the efforts of their various departments and their associated agencies to meet cross-cutting goals. Game fishes feature in a number of these initiatives.

The UK Government, for example, has set a suite of indicators of sustainable development with associated targets. These explicitly include the status of Atlantic salmon as both an indicator of ecosystem health and as an iconic species. The Department of Environment, Food and Rural Affairs (Defra) includes the 'Number of rivers in England with sustainable salmon stocks' as one of nine headline bio-diversity indicators, with the objective of protecting and enhancing Atlantic salmon stocks to ensure sustainable exploitation by fisheries as well as conservation of genetic diversity. The indicator relates to the number of rivers in which salmon spawning levels have met or exceeded agreed Conservation Limit standards for sustaining the native salmon stock. The Conservation Limit is set at a stock size below which further reductions in spawning numbers are likely to result in significant reductions in the numbers of juvenile fish produced in the next genera-tion, and hence returning adults from that year class. It also recognises that, as each river has a genetically distinct salmon stock, once the population of salmon is reduced below the number required to maintain a sustainable population, there is a significant risk of losing that genetic diversity. A target of 27 qualifying rivers, out of 40 rivers assessed annually, was set in 1999. Between 1997 and 2004, the number of rivers with sustainable salmon stocks varied between 13 and 25. Despite the significant bi-annual fluctuation pattern, more than 20 rivers have qualified between 2004 and 2008, and the target was exceeded in 2008 with 28 qualifying rivers identified. Despite this increase between 2004 and 2008, the number of qualifying rivers fell sharply in 2009 to 16, similar to numbers seen in 2001 and 2003. The overall trend in recent years is therefore tentatively upwards, taking account of the bi-annual fluctuation pattern. This is welcome, though no excuse for complacency, as 40 rivers meeting their Conservation Limit targets out of 40 surveyed is surely the only logical target if we seek a sustainable future in which

Fig 16.5 Building passes through barriers to migration is all part of restoring rivers for the benefit of their native complement of fishes.

critical aquatic and other environmental resources are not being systematically undermined by the cumulative pressures of contemporary lifestyles.

Likewise, management plans relating to river protection or enhancement, including those for rivers or river reaches designated under other legislation (including the EU Habitats Directive or as Sites of Special Scientific Interest under UK legislation), provide a vehicle to co-ordinate activities as diverse as development planning decisions, the targeting of agri-environment subsidies, river habitat restoration schemes and other such activities. The Environment Agency for England and Wales has also published strategies respectively for sea trout and salmon and also for trout and grayling, better to co-ordinate the efforts of its various departments and those of other government and non-government organisations.

16.5 Evolution of the common law

We also enjoy a twin-track legal system in the UK, mirrored in many other countries. Statute law sets down rules, whilst the common law evolves over time through case-law to protect people's rights. Protection of freshwater fisheries and benefits to wider water issues has produced a long and rich history of case law. This covers injunctions, damages and other forms of legal action associated with the attribution of 'property', both to fish in enclosed waters and fishing rights, and enjoyment of the angling experience more generally in fresh waters. The common law demonstrates a far greater flexibility than statute law, adapting to developments in knowledge, technology and changing patterns of enjoyment and utilisation of the water environment.

In simple terms, all people have a right of enjoyment of clean water and of various attributes of a healthy ecosystem. The establishment of the Anglers' Co-operative Association (ACA) in 1948 (subsequently renamed the Anglers' Conservation Association and later still reformed as Fish Legal in 2009) as a voluntary, membership-based organisation, deploying the common law to fight pollution and other harm to fishery ecosystems, is of itself significant in demonstrating the power of fisheries interests to promote the vitality of ecosystems delivering substantial public benefits. The first 'landmark' case taken by the ACA on behalf of an angling club was The Pride of Derby Angling Association Ltd v. British Celanese (1953), which established damage to the River Derwent fishery due to industrial pollution, subsequently resulting in substantial investment in sewage treatment infrastructure. Many novel cases were subsequently brought and won by the ACA, often establishing legal precedents and posing deterrents to would-be polluters. Two examples of this setting of precedents with respect to environmental pollution, both settled out of court, include Pool v. Gotto and Scott (1993), in which damages were recovered in recognition of the ecological effects of escaped rainbow trout, and Holmes, Collins and Hanley v. Critchlow and Critchlow (2002), which recovered damages for the diffuse effects of sheep dip pollution.

16.6 Cumulative evidence of changing rules

Legislative change may be slow, but there is a steady trace of change in the UK that has increasingly recognised and instituted protection for the many ways in which wider society benefits from ecosystems. This transition is dramatic when viewed in the compressed timescale of hindsight. For example, during the course of the twentieth century, British culture moved from one in which landowners had largely uncontested rights to use their land and natural resources as they saw fit, through to a legal and subsidy environment of increasingly stringent prohibitions and incentives relating to development planning, nature conservation, agricultural land use and river basin management for water resource and flood protection.

   This fundamental societal shift from private advantage to the recognition and the progressive and continuing safeguarding of the many benefits flowing to the

Fig 16.6 A healthy environment is becoming increasingly recognised as a central prerequisite of a healthy society.

wider population from landscapes, water and other natural resources, regardless of ownership status, is echoed around the already developed world. For example, questions have emerged about the longer-term and wider-scale impacts of major dams for ecosystems and human wellbeing, beyond the immediate advantages to influential sectors of society close to the dams themselves, and indeed the disproportionate influence of already privileged sectors of society on the management of natural resources of all kinds.

The world is changing, albeit slowly, progressively to embrace a wider conception of environmental quality, including thriving ecosystems and the benefits that they confer upon the breadth of society now and into the future.

CHAPTER 17

Changing values

T he vitality of wild populations of game fishes indicates both overall environmental health and the potential for ecosystems to support our needs into the future. Their current plight highlights the pressing need for rapid changes in the way we regard and use the natural world. However, change, particularly deep transformation of cultural practices and the overturning of long-established norms and vested interests, does not just happen instantly or by chance. It is instead a process of evolution wherein societal values shift progressively to reach 'tipping points', which change the norms of accepted practice, tradition and, in our Western culture, extension of the legislative and market bases. From changing public acceptance of driving whilst drunk or without wearing seat belts, the wearing of furs from rare animals, or the turning of a blind eye to child or bonded labour 'out of sight and out of mind' in the remote 'sweat shops' supplying our high streets, our ethics shift with increasing knowledge, media transparency and realisation of the consequences of our actions.

All of this calls to mind the pertinence of Aldo Leopold's call, published in his *Sand County Almanac* as long ago as 1949, for the expansion of human ethical codes, not merely to embrace selected strata of society but towards all of humanity and, by extension, to a broader 'land ethic' that values the ecosystems that support our diverse needs. The pressing requirement for a profound change in values and practices is beyond question. However, the pace of legislative change is in no way proportionate to the threats facing the modern world, nor can we expect law making to foment a culture change if not already substantiated by a groundswell of support that is either popular or shared amongst key influential stakeholders. Value systems therefore underpin all radical change, and it is to values that we must look to accelerate beyond the precautionary pace of legislative evolution.

17.1 Public concern for the environment

We have already explored the seeding of environmental consciousness from the later nineteenth century and its acceleration particularly since the 1960s, often

Fig 17.1 The twentieth century witnessed growing concern for the environment and recognition of its implications for human wellbeing.

with aquatic ecosystems and the wellbeing of dependent human communities playing a prominent role. This evolution of environmental literacy, deepening and broadening to embrace interdependent economic and social progress within our modern understanding of sustainable development, has at last begun to shape strategic decision making so that ecosystems are increasingly seen as mainstream concerns rather than mere afterthoughts. Although we undoubtedly still have a great deal of progress to make towards a truly sustainable world that fully safeguards its aquatic ecosystems and other crucial natural resources, a shift in societal concern for the environment is facilitating this journey.

This transition in perception of the intimate connection between the natural world and human security can be traced in transitions in humanitarian aid and international development. The emphasis of both these missions has shifted over recent decades, from relief 'flown in' from donor nations, often for only temporary relief, towards a contemporary emphasis on partnership working with local people to rebuild the resilience both of local communities and of their supporting environments in order to improve their ability to meet their own needs for the long-term future. This is extensively documented by the United Nations Development Programme (UNDP), particularly in its long-running (since 1994) annual series of *Human Development Reports*, which address different aspects of the international development agenda. These increasingly emphasise the close connection between human wellbeing and the supportive capacities of the natural environment.

In England, the government department Defra (and its predecessors) undertakes periodic surveys to develop a representative picture of what people in England think, and how they behave, addressing a range of issues relevant to the environment. These surveys cover parameters such as 'knowledge of and attitudes towards the environment', 'biodiversity' and 'green space'. Comparison between periodic surveys can provide trends in certain behaviours and attitudes (for example, the proportion of people who say they regularly recycle household waste). The 2011 publication of trends found that 92% of respondents said it was fairly or very important for them to have public gardens, parks, commons or other green spaces nearby, 48% said that they knew at least a little about biodiversity (increasing from 44% since 2009) and 78% when prompted agreed that they 'worry about changes to the countryside in the UK and loss of native animals and plants'. Furthermore, 13% had volunteered with, given time to or taken part in conservation volunteering for an organisation or community group in the previous twelve months. When prompted, 6 out of 10 people thought that river-water quality was a very important headline 'quality of life' issue, 87% were fairly or very worried about pollution in rivers, bathing waters and beaches, and a substantial majority also felt that biodiversity loss was a major concern, and something that they could act to change.

17.2 Business taking a lead

As 'the environment' broke through into cultural consciousness from the early 1970s onwards, governments around the world started the process of setting up departments with environmental responsibilities as well as bodies to regulate the conduct of other sectors of society. Formerly largely unconstrained by the impacts of primary resource demands, the environmental and ethical implications of supply chains, wastes from industrial processes and the fate of products, the understandable response of industry was commonly one of resistance to what was perceived as new constraints upon its freedom to operate through the imposition of 'unnecessary' restrictions and costs. In this early paradigm of environmental management, industry was often painted as the villain by environmental activists, with regulators playing the role of policemen, and government as an ambiguous champion of the interests of both.

However, we live in a world of progressively growing environmental awareness, which includes the dawning realisation and growing acknowledgement that sustainable development can represent a major business opportunity. Since the

1990s, a number of business sectors have taken a strong leadership role in the quest for more sustainable practices and associated new markets, seeing the enlightened self-interest that can stem from sustainable development of business models that better serve rather than undermine long-term societal interests. Exemplar business sectors in this regard include elements of the UK's PVC (polyvinyl chloride; plastic) industry, as told in Mark Everard's 2008 book *PVC: Reaching for Sustainability*, as well as some forestry, commercial fisheries and food businesses, amongst others, described by Mark Everard in his 2009 book *The Business of Biodiversity*.

There are some significant exemplars of business taking leadership, not merely in terms of overall environmental impact and innovation of processes and products, but of taking responsibility for stewardship of threatened natural resources. Forests are one of these dwindling ecosystems, with 30% of global land area disappearing to date under clear-felling, burning, flooding and degradation by conversion to commercial forestry and intensive agricultural practices at an unsustainable rate. Greenpeace, the campaigning environmental NGO, claims that 25 million acres of ancient forest are still being destroyed around the world every year.

This all has major implications for the water cycle and the vitality of aquatic resources, so it is heartening that the Forest Stewardship Council (FSC) was set up in 1994 by a consortium of interests that, at the time, seemed highly improbable. This grouping ranged from environmental NGOs (WWF, Friends of the Earth, Greenpeace), indigenous forest dwellers, professional forestry interests and major retailers such as Sweden's IKEA and the UK's B&Q. All shared a desire for a workable system that would promote responsible forest management practices and create a clear market for them. This bold, ground-breaking leadership created a market-oriented scheme well in advance of what national governments or inter-governmental groups could achieve.

FSC operates through independent and transparent accreditation to published standards, with the name 'Forest Stewardship Council', the acronym 'FSC' and the FSC logo all registered trademarks that can not be used without that prior authorisation. Certified products may carry the FSC label, including direct forest products such as timber, as well as products of a 'chain of custody' system, now including many processors, paper producers, printers and publishers, importers, retailers, architects, specifiers, self-builders and more product streams besides across the world. Thanks to the Forest Stewardship Council, an increasing proportion of the wood and other forest products used by industry and in consumer products originates in sustainably managed forests. As of December 2011, 147,831,804 hectares of forest were FSC-accredited through 1,078 certificates in 80 countries, taking it decisively out of a niche and strongly into the mainstream of the global market. This has significant, if as yet not adequately audited, implications for aquatic ecosystems and their fish populations and the wider environment.

Another exemplar market-led natural resource scheme directly relevant to fish stocks is the Marine Stewardship Council (MSC). MSC was co-founded in 1996 by WWF and Unilever (Unilever at the time purchased an estimated 20% of Europe's fish catches). The MSC was eventually constituted in 1999 as an independent, global not-for-profit body modelled largely on the successful processes of the FSC. The goal of the MSC is to achieve solutions to the problem of declining fish stocks with the key goal of contributing to the sustainability of ecosystems and fishery economies, which politically-based quota systems were manifestly failing to address. MSC accreditation is increasing the likelihood that fish consumed by people originates from a sustainable fishery, with a similar 'chain of custody' concept as used by the FSC to independently audit the transfer of fish from ecosystem to plate. It is in the interests of the billions of people dependent on fish for their protein source around the world, as well as the businesses that serve them and many others handling marine-based products, to halt or reverse sharply declining fish stocks.

The MSC's strategy has been to develop and promote a generic global standard for fisheries, backed up by independent certification and product labelling. Research underpinning the accrediting of a fishery includes an assessment of the fish stock and the impacts of fishing on the target population, including the methods used to catch them, as well as implications for the wider ecosystem. The MSC may certify either individual fisheries or bodies such as associations of fishermen or producers. If the certification is granted, fish or fish products from that fishery may carry the MSC logo. At this point, the 'chain of custody' from fishery through to retailer comes into effect and, if approved, the final consumer product can carry the MSC logo. Certification is a purely voluntary activity, but consumer pressure is considered to be creating impetus for increased industry involvement. In *Net Benefits*, the tenth anniversary publication of the MSC in 2009, the global MSC fisheries market was estimated to be over US$1.5 billion, with more than 2,500 individual MSC-labelled products available in 52 countries. This accounted for 150 fisheries engaged at some stage in the independent assessment process, representing 6 million tons of seafood, or about 7% of the annual global wild harvest. The MSC *Annual Report 2010/11* highlighted a 50% increase in the total number of fisheries certified to the MSC standard relative to the previous year, taking the fully certified total to 105, with nearly double the number of MSC-labelled products on sale around the world reaching close to 10,000 by the end of the financial year. Certified fisheries include Alaskan salmon, which are harvested based on the excess relative to calculated conservation needs, and which include such advanced technologies as spotter planes to monitor spawning fish escapement into river systems. The MSC's challenge now is to retain credibility by ensuring that the accredited fisheries remain sustainable, and by acting robustly by removing accreditation if monitoring shows they do not.

Fig 17.2 We desperately need sustainable fishing methods, by-catch of salmon and other fish by pelagic trawlers posing yet more threats.

There are a number of other market-based initiatives that, though not directly targeted at water protection, have significant implications for wider ecosystem health with positive consequences for aquatic ecosystems. The rise of organic agriculture is one such initiative, particularly as formalised since 1993 under EU Regulation as 'Organic' standards. These address foods and other farmed products, such as cotton, grown without most types of artificial fertiliser or pesticide, emphasising crop rotation, using natural fertilisers, protecting refuge habitat for the natural predators of crop pests, and maintaining the vitality of the soil. The rise of organic agriculture was driven by consumer concerns and the network of agricultural, wholesale, retail and catering businesses serving it, and therefore is a business-driven scheme with biodiversity and wider environmental health concerns at its core.

Aquaculture has also gone down the organic route, although whether this has been through a genuine desire to have less impact on the environment, or more of a marketing ploy, is a matter for conjecture. The Soil Association, the organic certification body in Britain, produced the criteria by which organic fish farms have to operate, but this does not completely stop the use of chemicals, for instance, in the control of parasites and some diseases. More to the point, the standards fail to fully address all the potential impacts from aquaculture on wild salmon and trout or the environment, and certainly fail to counter the perennial problem of the farming of predatory fish, such as salmon, that necessitates harvesting a far greater tonnage of sea fish to be fed (in pellet form) into cages than is finally produced in edible salmon or trout flesh.

One innovative example of how farmed fish were used to help their wild brethren was initiated on the Hampshire Avon in England from 1994. Under a deal brokered between the Wessex Salmon and Rivers Trust and the supermarket company Tesco Stores Ltd, then the largest fish retailer in the UK, Tesco agreed to reward every Avon angler voluntarily returning a fish with one of their farmed salmon. So the 'Tesco Swap a Salmon' scheme was born, under which, for every salmon caught on rod and line and returned alive to the water, Tesco donated a side of smoked farmed salmon to the captor. This contributed to the achievement of a 100% 'catch and release' record on the Hampshire Avon by the millennium. Tesco also provided core funding, which, with donations from members and some riparian owners, enabled the Wessex Salmon and Rivers Trust to fund the purchase of netted salmon for immediate live release. By 2006, the 'Tesco Swap a Salmon' scheme operated on all the southern and south-west rivers from the Itchen to the Fowey, with 'net and release' funding by then provided by The Avon and Stour Association. Some thousands of fish have survived as a result of these Wessex Salmon and Rivers Trust initiatives. The 100% 'catch and release' record on the Hampshire Avon still persists to this day, despite the cessation of the smoked salmon donations. In this way, spawning escapement was helped on a river that once boasted spring salmon averaging over 20 pounds, but now struggles to reach anywhere near its Conservation Limit of brood stock.

17.3 Fishing for better business

For a wide range of products, customers depend on the goods or services produced by biodiversity. Therefore, although the products themselves may not derive directly from raw biological resources, 'outdoor' and sporting goods manufacturers and countryside sports services are among businesses that already understand that they ultimately depend upon ecosystems for trade. Without fish for which to angle, diverse and beautiful landscapes in which to hike or camp, cycle through or otherwise enjoy, game to hunt, and scenery to paint or photograph, these businesses are not viable.

It is for this reason that some fishing tackle companies ring-fence a proportion of profit for reinvestment into the protection of vulnerable rivers and other waters. The Hardy and Greys fishing tackle company, based in the north-east of England, operates such a bursary scheme. The three strands of their 2007 programme included support for project work by the Wild Trout Trust and for the environmental campaigning work of both the Salmon & Trout Association and the North Atlantic Salmon Fund (all discussed in greater length elsewhere in this book). In addition to these three major organisations, Hardy and Greys also helps a number of smaller organisations on an ad hoc basis. Likewise, the Orvis fishing tackle and outdoor clothing company has supported such initiatives as the not-for-profit Riverfly Partnership's fundraising calendar for 2007, as well as work undertaken by the Wild Trout Trust and the Salmon & Trout Association. Between 2008 and 2011, the Fulling Mill Flies company donated all of their profits from retailing at the main UK Game Fair to the Salmon & Trout Association. Many other British fishing and fishery businesses and interests support, or have supported, organisations such as the ACA (Anglers' Conservation Association) and its successor Fish Legal.

In the USA, the not-for-profit organisation Trout Unlimited has been active in river conservation and rehabilitation, with the support of a wide membership that includes many businesses with associated interests. Likewise, the Wildlife and Sport Fish Restoration Program, run by the US Fish and Wildlife Service, is supported by various businesses, including, for example, a proportion of the profits from Stren Original fishing line. In a similar vein, many British shooting interests support the conservation works of the not-for-profit British Association for Shooting and Conservation (BASC), while, in the USA, waterfowling interests might support the not-for-profit Ducks Unlimited, which has had major successes in wetland creation and protection. Some outdoor pursuit clothing manufacturers support wildlife conservation initiatives for similar reasons. For this type of company, promotion of the ecosystems on which their customers depend or that they enjoy is not merely a 'good thing', it is also both a strategic corporate investment into a primary resource underpinning continued trade, as well as a differentiating feature in their marketing.

Lifestyle companies, too, have been active in their financial support for pro-conservation angling organisations. For example, Classic Malts of Scotland formerly co-sponsored the Wild Trout Trust. The Salmon & Trout Association has also previously enjoyed the sponsorship of whisky distilleries such as Laphroaig and Glenmorangie. The reason for this may partly be for the sponsoring business

Fig 17.3 Protection or restoration of wetlands can reinvigorate landscapes, helping fish and delivering many wider benefits to people.

to identify its brands with concerned anglers who are members of organisations promoting environmental protection for the benefit of fish stocks, but also because protection of salmon stocks in rivers automatically ensures the high-quality water required for distillery purposes.

Even the retail sector's economic giants, often pilloried for their heavy-handed relationships with suppliers, leading to adverse consequences for the livelihoods of farm businesses and for biodiversity due to downward pressure on farm gate prices, are increasingly seeking to identify with 'greener' consumers, building brand loyalty. For example, the massive retailer Tesco tries to gain customer attention and loyalty, such as through their nature-friendly flyer, *Little Steps to Becoming Greener*, distributed through UK stores in the summer of 2007. However, 'cause-related marketing' has both its strong champions and fierce critics. On the one hand, cause-related marketing, tying a brand to a social or environmental campaign, is a proven method for brand visibility and differentiation, as well as a generator of customer attention and loyalty. However, on the other hand, critics see cause-related marketing as giving some companies access to sectors of society on the back of shallow promises and/or less than generous returns to the charity or cause.

Ultimately, businesses will be judged on their depth of practical action by an increasingly sophisticated, environmentally literate and internet-enabled public seeking facts about the real impacts on the natural world, and no longer bamboozled by mere rhetoric. Today, all types of consumers are demanding greater accountability and responsibility from corporations. Set up and operated in a well thought through and transparent way, cause-related marketing can generate mutual 'wins' for the business, the cause and the sense of contribution to it by the customer. With appropriate commitments and safeguards, it is possible for positive steps for game fish stocks and wider biodiversity to ensue from businesses of all types through cause-related marketing, from which brand value and customer satisfaction also result.

17.4 A changing model of leadership

It is fascinating to watch this transition in innovation and leadership, taking responsible environmental stewardship out of the hands of the 'policemen' and into the 'enlightened self-interest' of business, which had been formerly perceived as amoral in its creation of wealth. The culture of business is indeed inherently more innovative and responsive to change than that of government. Prudent risk management is vital for any business that intends to stay around for any length of time, and particularly stewardship of its core resources, which extend well beyond financial and human resources to include the 'natural capital' supporting commercial interests.

Failure to address corporate impacts on water scarcity, aquatic ecosystems and myriad other natural resources is becoming increasingly self-evidently a symptom of poor risk management, demotivating staff and potentially attracting the ire of campaign groups and the media and, therefore, raising questions with investors. Neither does it impress ratings agencies or other stakeholders, and so it fails to safeguard corporate value. With some 70% of water use worldwide devoted to agriculture, and many other industry sectors dependent upon substantial withdrawals of fresh water, the value and attention that businesses place on vulnerable aquatic resources can make a substantial improvement to our world.

17.5 Living fish are worth more than dead fish

One of the many hard-won lessons from conservation programmes worldwide, from tigers, elephants and vultures in India through to medicinal plants in China and the Peace Parks movement in southern Africa, is that local people have to derive greater value from living organisms than they do from the quick meal or money resulting from killing them. This principle is as pertinent to fish as to other organisms, with large specimens of India's iconic mahseer fishes thriving largely where angling and ecotourism revenues enjoyed by local people exceed the benefits of killing them to eat or sell.

And so too for salmon. If one takes a commercially caught wild salmon of, say, 3 kilograms in weight, and assigns a generous price of £15 per kilogram to the commercial fisherman, then the netted fish is worth £45 at point of capture. Also, of course, the fish has to die to capture this financial return, and so is lost to the stock and the spawning escapement. However, if this same fish is caught by rod and line in the UK, it has at least a 60% chance of being returned alive to the water – and to spawn – while contributing anything from £500 to £2,000 to local economies, depending on the region in which it was caught.

And then one must look at the investment of fishery owners into river systems and fish conservation projects. A survey in 2005 on Hampshire's Rivers Test and

Fig 17.4 Recreational angling has provided a means for helping local people in the Himalayas derive more value from fish protection than from killing the fish using destructive methods.

Itchen showed that anglers spent £3.25 million to fish that year, of which £3 million was reinvested into management of those rivers by fishery owners supporting 120 full- and part-time jobs in the process. At least some of the other £0.25 million was spent in habitat rehabilitation projects by individual proprietors. As an over-all investment in the aquatic environment, these impressive values are not ex-ceeded by any other sector of society. And this does not include the money spent by those anglers in local pubs, hotels, bed and breakfasts and shops, which added greatly to those economies in areas where other employment was scarce.

The Scottish Executive's 2001 Green Paper, *Scotland's Freshwater Fish and Fisheries: Securing Their Future,* highlighted the lack of useful data quantifying the economic position of freshwater angling. This in turn led to the commissioning of a study of the impact of angler expenditure on output, income and employment on both a Scotland-wide and a regional basis. The outcome of this was the 2004 research report *The Economic Impact of Game and Coarse Angling in Scotland.* This research report estimated that freshwater fishing was worth £113 million annually to the Scottish economy, with salmon and sea trout anglers accounting for over 65% (£73 million) of this total. In addition, freshwater angling produced over £100 million worth of annual output for the Scottish economy, generating nearly £50 million in wages and self-employment income to Scottish households and

Fig 17.5 Fishery-driven investment in a buffer zone to improve a reach of the upper Bristol Avon was found by an Environment Agency study to deliver many broader benefits to the non-angling public.

supporting around 2,800 jobs. The report noted that this is a significant contribution, and that it should also be appreciated that salmon and sea trout angling has probably provided this level of annual contribution for most of the last century. Similarly, a survey in 2007 showed that angling on the River Tweed was worth more than £18 million to the Borders' economy, supporting around 500 full-time equivalent jobs.

South of the border again, two studies published by the Environment Agency in 2010 of the wide range of economic values resulting respectively from sea trout habitat restoration on the River Glaven in North Norfolk and the installation of a 'buffer zone' to protect a formerly severely cattle-poached field margin on the upper Bristol Avon in North Wiltshire, also revealed significant benefits to society and the wider economy. Though ostensibly focused on fish and angling interests, the River Glaven study recorded that angling benefits from the sea trout restoration project accounted for less than 1% of the total benefit of the project, which returned a substantial 325:1 benefit-to-cost ratio for the modest investment, with many benefits accruing from building social relations, stimulating ecotourism and regulating the environment, including a significant contribution to flood risk in the catchment. Likewise, the upper Bristol Avon buffer zone, though promoted by an angling club in collaboration with fishery officers of the Environment Agency, saw angling interests account for only 9.6% of the benefits of the scheme (still returning a favourable benefit-to-cost ratio of 3:1 to fishery interests) with the

Fig 17.6 An Environment Agency study of beneficial outcomes from restoring Norfolk's River Glaven demonstrates strong public concern and participation, and equally broad community-wide benefits.

remaining benefits (which including angling benefits returned a benefit-to-cost ratio of 31:1) accruing to wider society, including improved amenity, social relations, costs averted from nature conservation priorities, and erosion regulation.

All human activities have the potential to positively or negatively influence river catchments and their associated benefits to a diverse range of stakeholders. Therefore, an observed transition in the management of many recreational freshwater fisheries, abandoning the utilitarian maximisation of access and return per unit fishing effort in favour of recognition of the importance of working with natural processes and restoring river habitat and the natural regeneration of fish stocks, has a major contribution to make, not only to self-sustaining stocks of fish, but also to far broader dimensions of human wellbeing.

Ecosystem-based fishery management, if planned from the perspective of improving the functioning of river systems, can constitute a valuable 'building block', integrated with other localised river and catchment management initiatives, to contribute to restored functioning in entire river systems. By safeguarding or restoring ecosystem habitat, integrity and functioning, efforts expended on maintenance or regeneration of fishery ecosystems contribute to many benefits, often formerly overlooked, supporting the needs or enjoyment of wider sectors of society. Thus, if freshwater fishery management is integrated with other management initiatives as part of a wider vision of restored catchment functioning, they can make a significant contribution to societal wellbeing and economic progress beyond the angling community and its associated economic activities.

17.6 Religious leadership

The salmon is imbued with spiritual significance in some of the world's animist religions, including those of some native American tribes as well as Celtic religions, although not explicitly in the great religions, perhaps in part due to the absence of salmonid fishes from the Middle East and southern Asian heartlands in which they arose. Nevertheless, the symbol of the fish and the metaphor of 'fishers' is prominent in the Christian church, and this has been exported widely around the world by the predominantly Christian nations, which were first to industrialise and form global empires in the quest for new resources. Nevertheless, all religions have creeds commanding respect for the natural world and those with whom we share it.

Whilst the diverse religions of the world are often seen in the media as in conflict, the reality is that many common values bind them. From the sense of world citizenship and commitment to stewardship of the Earth under Baha'i and Zoroastrian faiths, emphasis of the interrelationships of all things under Buddhism, Shinto, and Sikhism, and the Hindu emphasis on the balance of creation, all inherently emphasise respect for all life, both human and natural, as an emanation of

Fig 17.7 Fish have long been associated with many religions and legends across the world and throughout history.

God. All in their own ways point to the frailties, weaknesses and the greed of humanity as the principal agents wreaking devastation upon creation, the need for us to improve ourselves, and our duties to others if we are to develop to live frugally and responsibly. With environmental responsibility and a regard for all people with whom we share the bounty of nature, we aspire to various conceptions of enlightenment or salvation.

However, beyond the metaphysical, the world's religions have also survived longer than any empire, monarchy, nation or company, and so many have embraced or encoded wisdoms pertinent to long-term stewardship. Stories of meaning are central to the teachings of Christianity, Islam, Hinduism and Buddhism. It is from our connectedness and dependence on the natural world, its seasons, storms, cycles and productivity, that early humans may have first derived the sense of a super-being and of codes of conduct that safeguard resources and their equitable sharing for the future.

Religions are also big. Two thirds of the global population are members of the world's eleven major faiths, more than one in six of the world population are Catholics, and 97% of the people of Papua New Guinea are Christians. The Christian church, Sikh temples and many other faiths are key partners in both local and international development. Furthermore, the holdings of the Church of England exceeded £4.8 billion in 2009. Many faiths have assets exceeding many banks or multinationals, and the collective shares of the major religions represent 6–8% of the global institutional investment market. Faiths own 7% of the habitable surface of the Earth, and have a role in 54% of schools.

The 'take-home' message here is that religions, regardless of our personal view of them, are rather better connected in terms of their teachings on environmental stewardship and our equitable sharing of nature than we may have considered. This is not to say that there is a 'grand unified theory' into which all religions fit; rather, their cumulative strength, like that of biodiversity itself, lies in the very pluralism of views and contexts from which creativity and adaptation may then arise. Various religions have been key partners in environmental, health and development initiatives across the globe, and many of their parables encode time-hewn wisdoms pertaining to environmentally responsible and ethical practice. Beyond this, through exercise of estate management, wise and long-term financial investment, outreach, education and other forms of influence, the value systems represented by the world's faiths have a huge and influential role to play in the attainment of a sustainable future that safeguards nature, including its game, iconic and other fishes, aquatic and other resources, and all who depend upon them.

Fig 17.8 People find meaning in salmonid fishes, symbolised here by 'The Pearl of the Dee', commissioned by the S&TA to celebrate the Queen Mother's 100th birthday. She wore it after it was presented to her, as did the Duchess of Cornwall when she started being seen publicly with the Prince of Wales.

The task is to unleash this massive power for good by contextualising threats to creation and humanity within our modern understanding of environmental and human conditions, and the urgent need – indeed moral duty – to strive to attain a sustainable pathway of development.

17.7 Cultural keystones

The Grand Coulee Dam in the US reach of the Columbia River in the Pacific Northwest region of North America, developed between 1933 and 1955 for hydro-power generation and irrigation supply, was a technological wonder of its age. Like many of the world's large dams, it was perceived as an unambiguous symbol of progress advancing the needs of all citizens. That was, of course, in an era of more blinkered thinking, focused on benefits accruing to influential sectors of society, and before the emergence of modern understanding of the collateral harm of massive dams for fish migration and recruitment, sediment flows in river systems and the starvation of habitats downstream, as well as impacts on water quality, flow regimes and, frequently, the proliferation of waterborne diseases. These negative consequences are, in practice, just a few amongst many wider, largely unanticipated adverse impacts of major dams. Today, this means that more dams are being decommissioned and removed in America than are being built, although the rate of dam building and creation of a legacy of catchment-wide problems continues at a staggering pace in India and China. These symbols of progress were famously described by India's first prime minister, Jawaharlal Nehru, as the '. . . temples of modern India', summarising their symbolic role to an increasingly technologically capable nation emerging in a post-war world. However, we now know that there are many casualties of this brave new technology, significantly including the native, ecologically and culturally important migratory fishes amongst many harmful effects on wider river systems yielding real and substantial human costs.

Fig 17.9 America's Grand Coulee Dam, a technological marvel with many associated problems (from the Library of Congress, Farm Security Administration – Office of War Information Photograph Collection Call number: LC-USW33-035035-C).

Where they occur, salmon and migratory trout suffer disproportionately beside other fish that run the river in order to reproduce. This certainly proved to be the case for the Grand Coulee Dam, effectively expunging the river's native salmon stocks and, with it, the livelihoods and culture of upstream Native Americans and Canadian First Nations, neither of which had been considered important in the rush to capture the dam's intended water storage and hydropower benefits. Production of salmon and other fish had been the centrepiece of the area's indigenous economy and culture. In 1951, the Colville Confederated Tribes filed a suit against the United States. Progress with litigation was, however, slow, requiring the playing out of the wider civil rights struggle to steer cultural values from those under which the privilege of ruling and economically influential classes became adequately diluted through recognition of the rights of all in society. Twenty-seven years after the claim had been filed, the Indian Claims Commission ruled in 1978 that the tribes were entitled to full compensation for all income losses associated with the dam. The US Government provided a massive US$66 million as historic compensation, including annual payments of US$15 million to offset ongoing reduced income opportunities. This huge compensation, and the protracted nature of the case, served to raise significant questions about the assumed advantages of large dams for all, including the

Fig 17.10 Hydropower turbines are far from a green solution (image courtesy of Phillip Ellis).

serious economic, cultural and environmental implications of extinguishing natural stocks of migratory fishes.

Though the cool, moist Atlantic climate of the British Isles has not driven us down the same route of massive dam building, the many smaller-scale dams, weirs, mills, barrages and other impoundments are not without their consequences for migratory fishes. (We have already considered the barrier to salmon represented by the ill-designed fish pass on the Tees Barrage in north-east England.) River impoundments are indeed associated with a rich British case law relating to damage of rights and enjoyment connected both with angling, commercial fisheries and wider uses of water and its loads of fertilising silt, progressively establishing precedents that advantages to a few local and often influential people should not outweigh the wellbeing of wider communities dependent upon the resources and processes of whole river systems.

Indeed, with renewable energy now of high political priority, applications for new hydropower schemes are set to soar, with forecasts of schemes being approved at thousands of river sites. While hydropower will certainly have its place in low-scale, carbon-neutral energy production, there is a crucial need for regulatory authorities (the Environment Agency in the case of England and Wales) to survey each catchment and decide where hydropower and efficient migratory fish passes can coexist to the benefit of both. There is an obvious potential for a 'death by a thousand cuts' effect from the cumulative effects of many, individually small schemes, through creation of barriers to migration

Fig 17.11 An Archimedean screw is hazardous for migrating fish, particularly where fish passes are lacking or inadequate.

or, almost as serious, slowing down migration, changing flow regimes, siltation and further simplification of habitat diversity in affected river reaches. The case for actual likely output of energy, taking account not merely of theoretical peak output but also of periods when turbines cannot be operated due to high flows as well as low water both of which are likely to be exacerbated under a changing climate, must be weighed objectively against cumulative impacts on fish of all species as well as wider beneficial services provided by river systems. The balance of benefits and costs across different stakeholder groups must also be explored dispassionately to ensure that decisions are equitable as well as environmentally justified.

Conversely, if hydroelectric schemes can be incorporated efficiently into existing weirs, and a suitable, efficient fish pass installed as part of the capital investment, then upstream fish movements could be eased and a 'win–win' situation achieved for local communities, low-carbon energy generation, economic gain and biodiversity together. There is no doubt that over-simplistic judgements about the 'greenness' of energy generation options could thwart progress towards sustainable 'win–win' outcomes, and that each scheme has to be considered on its merits, with the good ecological status of the local environment a prominent factor in the final determination. Indeed, in the light of European environmental legislation, this should be a given, although the rush for renewable energy all too often fails to see the potential environmental impacts in its wake, and the subsequent lack of 'green credentials', which will only become apparent in some future independent audit.

Fig 17.12 Our priority should be bypassing weirs that present impenetrable barriers to fish movement.

17.8 Icons of healthy waters

So, how do we know that our aquatic resources are healthy and capable of supporting diverse human needs? Historically, driven as we have been by an economic system born out of rapid commercialisation of resources, rather than their careful stewardship to assure indefinite future sustainability, the needs of nature have often been seen as in conflict with human needs in water allocation and other resource management decisions. This must change if we are to secure the basic environmental resources and processes which underwrite our future biophysical, economic and spiritual wellbeing. Increasingly, we need to see thriving populations of characteristic aquatic organisms, particularly iconic and economically important species such as salmon, trout and other characteristic fish species, not merely as 'nice to have' or conversely competing with societal demands for water, but as critical indicators of the vitality and supportive capacities of those ecosystems.

As we have already seen, the quality requirements of different types of fish population have formed the backbone of a great deal of water management in the UK, Europe and the USA over several decades. The needs of fish, explicitly including salmonid fisheries, have thus been long assumed and recognised to be important to the general health of aquatic environments. In this regard, game species are critical indicators of the health of those ecosystems upon which we now know we are dependent, and particularly for those species that depend upon networks of connected habitats, from upland spawning gravels through to food-rich riffles, into unimpeded lowland reaches, through estuaries and out to the open sea. Any breakage in this crucial chain – be that a physical obstacle, overfishing, pollution, siltation or other critical habitat modification, imbalance of predation, parasite transmission or other factors, or any combination of these – can pose serious threats to the viability of fish populations and the wellbeing of resources upon which they depend.

Fig 17.13 Salmonid fishes as icons of pure waters led to their inclusion in the logo of the National Rivers Authority (NRA), environmental regulator of English and Welsh waters from 1989–1996.

Fig 17.14 People care about the icons of clean waters, stimulating wonderful works of art like this painting of a brown trout by Robin Armstrong.

Our game fishes, then, serve as more than merely an indicator of healthy waters. Instead, they can be regarded as iconic of the ecosystems in which they occur. Some conservationists consider certain conspicuous and charismatic species to be 'flagship indicators', which mobilise public and political support for the conservation of the ecosystems necessary to sustain them. Other organisms are classified as 'keystone species', which serve as key ecological links characterising the ecosystems. Game fishes, particularly migratory game fishes, perform these roles but are also commercially valuable and culturally important resources in their own right, which may therefore shape local cultures and economies. For these prominent and generally large migratory fishes, including native British game species but also the mahseer fishes of India and the sturgeons of Eastern Europe and Central Asia, we assign the name 'iconic species', in recognition of the broad range of importance that they confer upon both ecosystems and dependent societies.

Fig 17.15 Native game fishes highlight healthy waters yielding many wider benefits to society.

17.9 Market connections

As we have seen, intact and functional ecosystems bestow many important beneficial services to humanity. Measures to protect or restore ecosystems may therefore have considerable value for wider society, over and above the direct benefits
stemming from the fish stocks themselves. However, since game fishes depend
upon diverse networks of habitats across whole catchments, migratory species
also moving into and through coastal seas and oceans, many of the people whose
activities positively benefit from fish stocks and wider ecosystem health may be
remote from key beneficiaries. A practical example here are upland farmers,
often eking a living on land of poor productivity, yet whose actions have a disproportionate effect on river hydrology, chemistry and the erosion of land, and the
successful reproduction of migratory game and other fish species spawning in or
close to headwaters.

This situation is unfortunate, albeit commonplace, but also raises issues of
equity when constraints are put on often already poor communities to reduce their
exploitation of these landscapes for the benefit of water users, fishing interests
and other constituencies downstream. Sustainability of fish stocks and the wider
environment, and equity in the way they are shared across society, are not only
ideals but also the stated goals of many nations. If we are serious about this, we
need to create mechanisms that place ecosystem integrity and the many ways in
which it underpins our wellbeing at the heart of planning and management, thus
linking ecological with economic and social progress. This entails linking 'producers'
of these services (those undertaking measures that address the protection or
enhancement of ecosystems) with their 'consumers,' including provision of water,
recreation, flood protection, waste assimilation, culturally-valued landscapes and
other services to settlements and businesses downstream.

One of the effective and potentially sustainable ways that this has been achieved
is through the creation of markets to link up these players. There are emerging
initiatives around the world that implement a 'paying for ecosystem services'
(PES) approach, founded on the creation of markets connecting 'providers' with
'users'. Many examples of evolving schemes are found in South Africa, where
innovative water laws instituted as the country emerged from apartheid into
democracy during the late 1990s enshrine the principles of equity, sustainability
and efficiency. This has promoted a range of pioneering approaches to adaptive
governance, accounting for implications for the range of ecosystem services
in development decisions, and detailed economic studies of the market value
delivered by catchments, together with the marginal implications of different
development options.

One such pertinent initiative is that in the Maloti Drakensberg, which recognises
that water yield from the cool, mountainous uplands of Lesotho and the high

Drakensberg in South Africa are essential water sources for often remote users in the more productive, yet arid, plains of lowland areas. A study in 2007 by the Maloti Drakensberg Transfrontier Project seeks linkages between the restoration and management of upper catchment areas for the purposes of increasing run-off of water, and those who benefit economically from improved water quality and quantity lower in river systems. This project also builds on other progressive initiatives already in place in the region, with the intent of establishing a market mechanism by which beneficiaries of heavy water use downstream (forestry, intensive agriculture, particularly sugar production, mining, industries such as paper mills, etc.) can invest in work to increase the water yield of the catchments upon which they depend, rewarding and providing a stable income for often poorer upland communities required to modify their practices. This market model has found favour with the South African Government, which hopes to further develop such PES markets, both here and across the country, founded on the natural functions of catchment ecosystems as a basis for the equitable, sustainable and efficient provision of water.

Three further exemplars relate to restoring the natural functions of the water environment, with significant water resource benefits.

Firstly, the public water supply to New York City provides a dramatic illustration of the economic value of natural, ecosystem-mediated water storage and purification processes. NYC's Department of Environmental Protection delivers over 1.2 billion US gallons (4.5 billion litres) of water daily to 9 million people within an infrastructure that has evolved with the city, from local wells and smaller water bodies into one that draws from around 2,000 square miles (830,000 hectares) across the wider Catskills and Delaware catchments to the north. Traditionally, the city had relied on sparse populations and low-intensity land use across these rural catchments yet, by the 1980s, industrial-scale agriculture was replacing traditional methods and residential development added to the threats in these environmentally vulnerable areas. At the same time, public health standards were becoming more stringent, potentially posing the city with bills of billions, or tens of billions, of dollars for new filtration plant. However, cost-benefit analysis suggested that a comprehensive programme of watershed protection would be substantially more efficient than the downstream filtration of contaminated water. This set city authorities on a tortuous pathway of negotiating, through patient dialogue and consensus building, a rural–urban partnership of mutual benefits to farmers and water users across New York. By the end of 1991, the city and the farmers had begun to implement an urban–rural watershed protection partnership, entailing funding of whole farm plans that integrated agricultural pollution control into individual, economically efficient farm business plans and grants for associated capital works. At a total cost of approximately $1.3 billion (£700 million), the watershed protection programme will maintain New York City's pristine water

Fig 17.16 Healthy fish populations indicate water in good condition for human and other uses, with associated economic benefits.

quality for the foreseeable future, to the advantage of communities and also to the vitality of the freshwater ecosystems and game and other fish populations of the wider region.

The second exemplar is SCaMP, the Sustainable Catchment Management Programme, initiated by United Utilities (the major provider of water and sewerage services in the north-west of England) in association with the RSPB, from around 2002. United Utilities owns substantial areas of uplands important as water-collection zones serving its infrastructure and the supply to millions of household and business users primarily in the lowlands. SCaMP recognises the central role of ecosystems in producing high-quality water, and has entailed funding of habitat restoration to boost both moorland biodiversity and yields of water of higher quality and more dependable flows. A five-year review of monitoring outcomes from the SCaMP programme published in 2011 found that it is working towards delivering on its objectives though noting that there is a long lag phase in the recovery of ecosystems and their functioning phase, also potentially contributing towards flood alleviation downstream and to the recruitment of fish and the vitality of wider river ecosystems. By restoring habitat for biodiversity, the uplands of north-west England benefit from improved hydrology, water quality, landscape and amenity, security of tenant farm incomes, and retention of the character, resilience and economic viability of traditional landscapes.

The third example is that of the multi-billion US dollar Vittel bottled-water business. Water marketed under the Vittel label is drawn from the 'Grande Source' ('Great Spring') located in the town of Vittel at the foot of the Vosges Mountains in north eastern France, to which beneficial properties have been ascribed since Gallo-Roman times. Yet rising nitrate concentrations in the Grande Source, linked to intensification of agriculture in the groundwater catchment, were posing a looming threat by the mid-1980s. This led the then family owners of the spring (the Vittel brand is now part of the Nestlé Waters group) to assess a number of approaches to ensuring water quality. Of these, the only viable approach was to provide incentives to farmers to voluntarily change their practices. This raised a substantial challenge in making Vittel and farmers' interests coincide such that it was in the interests of farmers to co-operate. This initiated a long process, entailing the setting up of a multidisciplinary research partnership and also an independent broker organisation, that eventually secured the co-operation of a key group of farmers with the greatest influence on the aquifer. Through a system of incentives paid for by the key commercial beneficiary of clean spring water, farmers in the Vittel catchment have been enabled profitably to revert to extensive farming systems, simultaneously addressing land, labour, capital shortage and long-term agreements offering security of farm income, farm conversion and source protection. The Vittel case study included a clear market between one buyer (now Nestlé Waters) and 26 voluntary 'sellers' (participating farmers) likely to make the greatest difference to water quality based on a scientific rationale, itself centred on a 'learning by doing' approach.

Further examples of PES schemes are now evolving across the UK and Europe, east and southern Africa, Australia, south and central America and elsewhere across the globe, addressing services ranging from water supply to carbon sequestration, erosion control and landscape rehabilitation, nature conservation and flood resilience. They are also implicit in changing attitudes to the outputs of land use in the 2003 revision of the European Union's Common Agricultural Policy (CAP), which shifts focus on subsidies from production of commodities onto agri-environment support intended to provide wider societal benefits that may include heritage, biodiversity and a variety of other facets of ecosystem functioning. Broadening the concept of farming from production of commodities alone into a model that recognises the value of water production, landscape and other ecosystem services, puts us on an inherently more sustainable pathway in which the vitality of supporting ecosystems, including their game and other fish populations, are recognised and progressively more deeply embedded in markets for their capacities to support a diversity of human needs. Hopefully, a further review of CAP in 2013 will further strengthen support for realigning agricultural subsidies with environmental protection that delivers wider societal benefits.

17.10 The economic prudence of an ecosystems approach

Our fragmented model of environmental management in the past has often confronted us with major headaches in prioritising expenditure, for example on abatement of water pollution versus addressing nature conservation targets, protecting culturally important landscapes, controlling flooding, or building infrastructure to retain water resources. There have also been frequent conflicts between environmental outcomes in the way we have made these investments in the past, for example the massive increase in emissions of climate-active gases associated with advanced treatment of wastewater in the last two decades of the twentieth century. All too often, from this blinkered view of perceived disconnected environmental problems, fragmented management solutions have produced further unintended consequences of their own. Of course, resource constraints will always apply, but we have to consider looking at the environment in a different and far more integrated way.

Much of our legacy of environmental management practices has addressed the pressures we place upon the environment and the measures we need to take to reduce them. Thereby, we implicitly assume a right to develop land or devise technologies without their interaction with ecosystems framing the direction of development, only retrospectively considering measures to reduce or mitigate their consequences for ecosystems and people. So we have seen, for example, extensive drainage of land and physical flood banks built to stimulate food production in the post-Second World War period, and only later have we realised the impacts of this on exacerbation of flooding elsewhere, loss of habitats and wildlife, negative impacts on fish recruitment and valued landscapes, reduction in environmental purification processes, and a range of other problems that require

Fig 17.17 Vibrant watercourses and ecosystems provide many often undervalued benefits to society.

us to invest precious resources to redress unintended environmental damage and costs to society. With hindsight, we are today embarking on a strategy to set back flood defences and rewet land such that it can naturally absorb flood surges and function in many ways for the benefit of nature and society, and are also beginning to reflect on our lack of foresight in effectively spending ten pounds to deliver a pound of food production value. As discussed above, the challenge is now to ensure that we do not make the same mistakes with low head hydropower development, especially when the evidence is staring decision makers in the face as to the potential for broader environmental damage to which a blind eye may be turned only at substantial future risk and potential liability.

Today, we are learning, informed particularly by the rise of ecosystem services as a tool to guide policy and practice, to consider the environment not merely as a receptor of our pressures but as a living resource to be safeguarded and valued for its capacity to support our multiple and diverse needs and interests. Salmon, trout and other native fishes are invaluable in this regard, their iconic status helping us recognise that intact ecosystems capable of supporting valued and characteristic species deliver a diversity of interconnected societal benefits effectively 'for free'. In this regard, the two Environment Agency studies noted above (the River Glaven and the Bristol Avon) are particularly significant, addressing and valuing the outcomes from ecosystem restoration initiatives which, though ostensibly fisheries driven, were actually found to deliver an overwhelming bulk of benefits to sectors of society and the economy significantly beyond fisheries interests.

We will always suffer resource constraints in addressing environmental problems. However, by placing the vitality and supportive capacities of ecosystems centrally in our considerations, we may not only better protect natural resources fundamental to our future wellbeing, but also do so in a way that spends a pound to deliver ten pounds of social value.

CHAPTER 18

We the people

There is just something not quite right about a river devoid of its natural complement of fish or, indeed, devoid of fish at all. It is like a summer without swallows and martins swooping low over waters and meadows to intercept flies, or with skies silenced of the screech of chasing swifts, an autumn robbed of the blaze of turning leaves, or dawn without a crescendo of songbirds.

We can make compelling technical cases for the ecological 'wrongness' of these things but, over and above this, we simply feel it viscerally, regardless of our rational understanding. So, whose job is it to express these feelings and rational arguments, and to create a groundswell of activism that results in changing perception, policy and practice?

18.1 People power

It is not about what governments, businesses and religions think in some auto-cratic, top-down sense. Ultimately, it is people that make the difference. People care about their futures and those of their loved ones. Ultimately, it is people who get concerned about Rachel Carson's bleak imagining of a *Silent Spring* or of the prospects of a barren river, and their articulation of this is the potential rootstock of cultural change. However that concern subsequently manifests, whether through the organisation of campaigns and campaigning groups, in religious or political conviction or economic implications, when people get together to express concerted opinions or to take practical action, it is 'people power' that changes the world. Democratic change is not possible without the expressed opinions of people.

In this regard, the iconic role of the game fishes may be of particular signific-ance. Historic class issues may mean that some of the angling and wider public have little regard for salmon and trout and their protection. However, for the majority of people, the very presence of these fishes has significance. Indeed, a survey in 2006 by the Environment Agency showed that the people of England and Wales would be prepared for £350 million per annum of government money to be spent on ensuring salmon still ran their rivers; an astonishing 'existence value' that had nothing to do with the catching of fish. Furthermore, as we have seen in considering the many 'ecosystem services' that support our needs, all in society ultimately benefit from ecosystems healthy and functional enough to support these fishes.

We have already considered the origins of the environmental movement through such bodies as the RSPB, and the mobilisation of angling interests to protect the vitality of fish stocks though organisations such as the ACA. However, many other civil organisations across the UK and around the world have the protection of fresh waters and fish stocks at or close to their heart, some of them actively restoring rivers and fisheries.

18.2 Champions of the game fishes

In much the same way, and at much the same time, as the formation of what was to become the RSPB, representing public concerns about birds, a parallel body was forming to champion the interests of salmon and trout in the UK. The Salmon & Trout Association (the S&TA) was formed in 1903 over concerns for the state of our rivers from the impacts of the Industrial Revolution. In those early days, there was as much concern for the future of commercial salmon netting as for the fish themselves, but the association gradually evolved during the twentieth century to become one of the most respected repositories of knowledge about game fishes in the UK. The association's magazine also published many scientific papers between its inception in 1910 and the 1980s.

A great deal of change has occurred in the wider world since 1903, and the S&TA has had to evolve to keep its work relevant to the current issues affecting the management and conservation of game fishes and the environments necessary for them to thrive. By the 1980s, the S&TA's supporters were mainly from the ranks of fishery owners and game anglers, although the priority was still the wellbeing of the fish ahead of any other consideration. This culminated in the granting of charitable status to the association in 2008; recognition that the S&TA's work over more than a century had indeed been of wider benefit than for just its immediate target angling audience.

This transition into a charity reflects the extent to which the S&TA's campaigning and other activities, addressing all issues relevant to fisheries legislation and regulation as well as environmental and species management and conservation, promotes the many wider public benefits that flow from waters fit for their natural complement of salmon, trout and other characteristic fishes. To achieve this, the association has developed close working relationships with government departments and agencies, including contacts in both houses of parliament, the devolved UK Governments and with senior officials in the European Commission.

Fig 18.1 The Salmon & Trout Association, long-term champions of salmonid fishes and now constituted as a charity '. . . for fish, people, the environment'.

It advises them over fisheries and angling matters and works to influence their decision-making processes on behalf of all those with an interest in the aquatic environment.

Most importantly, the S&TA now supports all its policies and arguments with peer-reviewed scientific evidence and legal opinion, elevating its role considerably beyond that of a single-interest angling group. Through the vehicle of game fishes, therefore, and substantially supported by those that like to fish for them, the S&TA works on behalf of all who share an interest, whatever that interest may be, in the management and conservation of the aquatic environment and its dependent species. Furthermore, the association backs up this role with a substantial educational programme aimed at all ages and social backgrounds. This ranges from beginners' fly-fishing courses, which include an introduction to fly life identification and other environmental issues, to the publication and distribution of schools' literature on the life-cycle of game fishes and the different challenges they face in an environment managed by humans. Inspiring the next generation to take up the stewardship role in the aquatic environment is an essential objective for the S&TA.

Another voluntary body championing the interests of the Atlantic salmon is the Atlantic Salmon Trust (AST), a UK-based organisation with Atlantic-wide interests dedicated to championing wild salmon and sea trout, particularly during the marine phase of their life-cycles. Like the S&TA, the AST does not represent specific interests or bodies, focusing its efforts on the conservation and improvement of wild salmon and sea trout stocks to a level that allows for their sustainable exploitation. The Trust was set up in 1967 in the wake of the then recent and disastrous UDN (ulcerative dermal necrosis) disease outbreak, to address widespread concerns about the decline of stocks of these migratory fishes, as well as to conduct and support practical research into specific problems in marine and freshwater environments. AST also gives practical advice on the management of fisheries and rivers, as well as offering scientific advice to governments, international and national authorities and commercial enterprises.

The UK, then, has not one but two politically-active and science-based organisations channelling public concern for declining game fish populations into positive influence and sound advice to political and other audiences.

18.3 Rivers in Trust

Fig 18.2 The Wye and Usk Foundation have the emblem of all life stages of the Atlantic salmon as their logo, and also a focus for their effective charitable work.

Another dramatic demonstration of the mobilisation of people to achieve beneficial outcomes for game and other fish species iconic of the broader ecosystems that support them, is seen in the rise of the river Trust movement in the UK since the 1990s. The various Trusts, established on river systems from England and Wales and into Scotland, have commonly been initiated on the basis of fisheries interests, with particular concern for game-fish populations leading the way. However, all of them have broadened rapidly to reflect interests in nature conservation, preservation of heritage, security of rural incomes and a range of other factors associated with the vitality of river and catchment systems. The charitable objects of the Wye and Usk Foundation (WUF), for example, relate explicitly to the regeneration of Atlantic salmon stocks in these two river systems.

Likewise, the Westcountry Rivers Trust, as well as the Eden Rivers Trust and the Tweed Foundation, were instigated by trustees interested in the protection and regeneration of natural salmon fisheries, but focus now on the vitality of the ecosystems that support not only populations of these and other game fish species but also a broad range of ecological, cultural, economic and other benefits associated with them.

The primary emphasis of the river Trust movement is upon practical projects to regenerate the ecosystems of local rivers. This includes a focus on measures to restore the damaged features and ecology of the whole catchment, from its headwaters to the sea, as well as campaigning against excessive high seas netting of adult salmon and sea trout that would otherwise return to their natal rivers. Many of the Trusts, from small organisations such as the Wandle Trust in south London, through to regional-scale bodies such as the Westcountry Rivers Trust, operate excellent educational programmes. These range from public meetings and campaigns down to

Fig 18.3 The Westcountry Rivers Trust works to protect the whole river ecosystem and its many human benefits, the trout in their logo together with the otter and mayfly reflecting the interconnections in the system.

Fig 18.4 Volunteers regularly muck in on Wandle Trust clean-up days on this south London river (photo courtesy of the Wandle Trust).

classroom-scale initiatives, such as the innovative 'trout in the classroom' programme, which provides fish tanks and trout eggs to local primary schools so that children can watch and learn from the life-cycle of these fishes, to the point that 'their fish' are released into local streams engendering further concern from local families for their continued welfare.

Many of the river Trusts have been successful in winning substantial regional development funding from the EU, businesses particularly including water service companies, environmental regulators, government departments and other sources on the basis that regeneration of the river can be a focus for improving regional prosperity and better quality natural resources, including diversifying and stabilising farm incomes, promoting regional ecotourism including angling, enhancing the quality and reliability of the source of water used for abstraction to public supply, and other means.

One of the many innovations of the river Trust movement stemmed from the conundrum of how to maintain self-interest for farm businesses in order that they continued to manage their riparian land positively after advisory visits and capital funding routed through regional Trusts had ceased. The solution was the instigation of 'Angling Passport' schemes, under which anglers buy booklets of tickets to fish listed waters, signing and posting tickets into collection boxes before fishing, with fishery owners then passing these tickets back to the Trust to receive payment for angling visits. A 'virtuous circle' is created, whereby anglers benefit from flexible access to new waters with, generally, native self-sustaining fish stocks, whilst simultaneously riparian owners receive revenues from angling, often in restored river reaches where angling had not occurred before. This secures a vested interest for them to continue to protect and enhance the interface between their land and the waterways that traverse it. The net beneficiaries of restored river reaches and more sensitively farmed riparian land, and their continued maintenance, includes not only the fish stocks and the ecosystems themselves, as well as the anglers that exploit them, but also broader constituencies of beneficiaries of the many ecosystem services that they provide. Economic valuation studies on the wider consequences of such river restoration programmes reveal substantial benefits beyond those forming the primary targets of the initiatives, also including carbon sequestration, improved regulation of flooding and agricultural pests, landscape enhancement and a range of other benefits exceeding many times investment in these schemes.

Atlantic salmon were also the explicit focus of Trust initiatives in the River Thames. Improvement of the environment of the Thames had taken significant strides forward since the 'great stink' of the 1850s, which saw epidemics in London of cholera and a wide range of other diseases that we now know to be waterborne, also precipitating the suspension of Parliament in 1858. This was a spur to government to face up to the need for a more wide-ranging urban planning policy, and for the commissioning of a system of sewers to intercept sewage from entering the Thames in London, diverting it instead to new treatment works in the East End. Further progress was made over the ensuing century. However, July 1986 saw the constitution of the Thames Salmon Trust as a registered charity, with the ambitious aim of bringing about regeneration of the river such that Atlantic salmon would again be able to run the Thames. The organisation was reconstituted in 2005 as the Thames Rivers Restoration Trust, with the broader objectives of conservation, protection and rehabilitation of the habitat and waters of the Thames catchment for the benefit of all indigenous

Fig 18.5 The Rivers Trust's 'Angling Passport' scheme, creating incentives for landowners to maintain healthy rivers.

species, including Atlantic salmon and migratory trout. The Thames Rivers Restoration Trust essentially co-ordinates the limited investments of environmental regulators, water companies, concerned businesses and other bodies operating across the Thames to improve water quality, passage through weirs and other obstructions, and otherwise promote river ecology, including the opportunity for natural recolonisation of the river system by migratory salmon and trout. Atlantic salmon have returned, albeit as yet in very small numbers, which is in many ways a miracle given the noxious state of the river a century and more ago, and so a cause for hope. Plans by Thames Water, the regional water services company, to build a new storm overflow pipe to intercept sewage-contaminated water in heavy rain conditions and carry it away from the river to storage and processing facilities could prove the next stage in cleaning up the tidal stretch through the City of London. In many ways, this new venture echoes the great sewer-building initiative instigated in the late 1850s. Removal of the regular pollution barrier moving up and down on the tide might well encourage returning migratory salmon and trout to run the river in sufficient numbers to create self-sustaining populations of Thames salmon and sea trout. We live in hope, but will have to wait some years to see if this grand vision will be achieved.

More widely across the British Isles, a significant change is taking place in river management. Formerly, many management decisions and actions were focused on local priorities for flood defence, fishery performance and water quality control based on point source discharges such as sewage or industrial effluent. Today, scientific awareness is helping us develop a more connected, systemic perspective on how rivers function, and how better to manage habitat, land use and the various pressures we place on river systems, in a more integrated way. A major champion of this transition, together with the bodies already mentioned, is the Wild Trout Trust, which is a practical body working with landowners to restore the natural functioning of rivers for their desirable, economically valuable and biologically important, self-sustaining fish stocks. As the Wild Trout Trust's name suggests, the focus is particularly on naturally spawned migratory and non-migratory trout species, though other fishes, too, benefit from more healthy river systems.

All Wild Trout Trust work is based on a similar model of working to overcome 'bottlenecks' to the natural recruitment and production of native fish in rivers, requiring a change in paradigm of fisheries management that emphasises restoration of river habitat, water quality, hydrological regimes and ecosystem intactness. Associated management measures may include rehabilitation of spawning gravels and other essential habitats, installation of 'buffer zones' or farm advice to abate diffuse pollution, reporting and lobbying to reduce point source pollution, restoring natural flow regimes or compensating for over-abstraction by techniques such as channel narrowing or installation of deflectors to improve local channel scour,

Fig 18.6 The Wild Trout Trust works for the recovery of wild, self-sustaining brown trout populations, the trout on their logo representing thriving waters yielding many benefits to people.

and control of invasive species of plants and animals compromising ecosystem integrity. All of these measures focus on the integrity and optimal functioning of natural ecosystem processes, which will have inevitable co-benefits for many other species and people throughout restored river catchments. This is especially relevant to Wild Trout Trust projects working to ensure that trout can thrive within river stretches through towns and cities, accessible to people and representing part of the wide range of benefits provided by healthy 'blue infrastructure'.

It is not just river Trusts that harness the positive aspirations of those with an interest in thriving fish populations. The network of county wildlife Trusts, too, frequently see fish, particularly those such as grayling and trout that require clean environments, as indicators of successful reserve management. A practical example here is the River Wylye running through

Fig 18.7 Live willow deflectors installed in a formerly straightened stretch of river, restoring much-needed flow diversity, scour, nursery and flood/predator refuge habitat.

Fig 18.8 'Buffer zone' installation to keep stock out of rivers and allow habitat regeneration can make a major contribution to restoration of rivers and their fish stocks.

the Langford Lakes wildlife reserve, operated by the Wiltshire Wildlife Trust, in which grayling are not only a target species for river management but also the centrepiece of the recreational fishery operated by the Trust to help finance management of the lakes and other ecosystems across the reserve.

The successes of the river Trusts include not merely enhancing river systems for the benefits of game and other fishes, wildlife, local people and the local economy, but in demonstrating that putting ecosystems at the heart of management programmes yields significant and enduring benefits to wider sectors of society. Salmon, trout and other fishes again form a focus around which these many broader societal benefits are marshalled.

18.4 Anglers in Trust

The concerns of anglers, not merely for their sport but from a deeper place that compels us to gravitate to living waters, have been prominent in the promotion of advocacy, research, education, practical restoration and general mobilisation of society to establish and achieve the outcomes of the voluntary organisations highlighted in this chapter. After all, the thrill of hooking a fish skims only lightly on the surface of motivations that have led the angling community to be among the most strident and active champions of the wellbeing of river systems. The capture of fish itself is also but a tiny part of the voluminous angling literature that has accumulated over many centuries.

Outside of the arena of angling, many people who enjoy pursuing game fishes are also active in other voluntary, scientific, policy-making, business, education, environmental management and nature-related fields. We are all, in our unique ways, potential campaigners and champions for the wellbeing of our rivers and their fish fauna and wider ecosystems, all of us with a sphere of influence into which we can promote the slow transition of society from net consumer to wise steward of the ecosystems that support our biological, economic and quality of life needs into the future.

The Riverfly Partnership is a valuable expression of this groundswell of public interest in the health of British rivers, particularly amongst the fly-angling community but certainly far from exclusively so. As the name suggests, the Riverfly Partnership focuses on the fly life of British rivers. As indicators of the overall health of our waters, emerging fly life, including flies of interest to the fly angler, are important and visible elements of healthy rivers. A substantial body of corroborating anecdotal evidence has pointed to a general 'malaise' in the vitality of trout streams in the latter part of the twentieth and the early twenty-first centuries, highlighted by declining hatches of many characteristic riverfly species. The Riverfly Partnership, hosted by the Salmon & Trout Association, is a network of nearly a hundred partner organisations representing anglers, conservationists, entomologists, scientists, watercourse managers and relevant authorities who share a concern and wish to do something positive about the situation. These

Fig 18.9 The Riverfly Partnership trains and involves volunteers in riverfly identification, monitoring, education and conservation activities.

disparate bodies work together to protect the water quality of our rivers, further understand riverfly populations, and actively conserve riverfly habitats. The partnership's diverse activities include providing a forum for raising issues affecting riverflies, increasing awareness of their conservation importance, involving people in monitoring and recording, stimulating research, and working to improve conservation status. Active volunteer involvement is central to the wider outreach and mission of the Riverfly Partnership.

18.5 The iconic status of salmon and other game fishes

Our industrial history and the economic system we have inherited from it are predicated on the rapid commercialisation of resources, rather than their careful stewardship to assure indefinite future sustainability. This has often posited the needs of nature as in conflict with human demands for water and other resources. This must change if we are to secure the basic environmental resources and processes which underwrite our future biophysical, economic and spiritual wellbeing. Increasingly, we need to see thriving populations of characteristic aquatic organisms, particularly characteristic and economically important species such as salmon and trout, not merely as 'nice to have' or conversely competing with societal demands for water, but as critical indicators of the vitality and supportive capacities of those ecosystems.

As we have already seen, the quality requirements of different types of fish population have formed the backbone of a great deal of water management in the UK, Europe and the USA over several decades. The needs of fish, explicitly including salmonid fisheries, have thus been long assumed and recognised to be important to the general health of aquatic environments.

Other terrestrial environments have their 'charismatic megafauna'. Large and popular birds in the British landscape, along with the tiger and elephant in India and the giant panda in China, are often referred to as 'flagship species'. This is because their survival depends upon the conservation of networks of habitats required to support their whole life-cycles, and they also attract a great deal of public affection and support, making them powerful levers for conservation action, even though other less charismatic species may play more important roles in these same ecosystems.

Salmon and sea trout also share these attributes, requiring networks of connected habitats, including, in the case of Atlantic salmon, clean upland headwaters where eggs and fry are laid and nurtured, parr gradually moving down the river system until smolts occupy estuarine reaches and eventually take to sea, with the maturing salmon then living as marine predators in the open ocean until their eventual return to estuaries for the subsequent run up to the headwaters in which they were spawned. Salmon are therefore vulnerable across this whole interconnected set of environments, including high seas, coastal and estuarial netting, trapping in estuaries, impoundments and other physical obstacles across rivers, pollution, habitat degradation, diseases including those arising from cumulative stresses, over-abstraction leading to declining river flows, siltation of spawning and nursery gravels, over-exploitation or poaching, and potential predation. All of these pressures represent potential breaks in a chain that could threaten the continued recruitment of salmon populations. The conservation of Atlantic salmon therefore depends upon sensitive management of this whole complex of habitats, and the transitions and connections between them, including and also benefiting the many other species that share it.

Migratory species such as salmon and trout go beyond 'flagship' conservation status alone. They, along with other large migratory fishes of the world such as the sturgeons of Eastern Europe, the mahseer of southern Asia and the somewhat smaller but no less habitat-dependent yellowfishes of southern Africa, also have an economic and cultural importance that includes value accruing to, and sometimes even defining, the livelihoods of local communities from their exploitation for subsistence, commercial and/or recreational purposes. Successful Atlantic salmon populations, therefore, not only indicate healthy networks of habitats supporting different phases of their life-cycles, but also have significant broader societal importance, and for this they have been ascribed the term 'iconic species'.

Fig 18.10 Fish and water symbology is everywhere in society, as here on the crest of the Fishmongers' Company, one of the twelve Great City Livery Companies in London.

It is true that freshwater fish do not generally arouse great emotion in the public, or have an appeal equivalent to birds (note that the RSPB has in excess of one million members in the UK), certain iconic fish species, nevertheless, have been proven to have substantial broader public resonance. The Thames Salmon Trust proved a successful focal point for the energies of regulatory and voluntary bodies from its inception in 1986, not due to widespread interest in angling in a river that at the time was devoid of salmon, but because the Atlantic salmon itself was the most appropriate icon of a river restored to full health, with all of the many societal benefits that flow from that goal. The very presence of this iconic fish was widely conceived as an unambiguous symbol of the river rising, phoenix-like, from years of neglect and abuse. Significant press profile has rightly and understandably greeted the return of salmon to rivers such as the Mersey and the Don, formerly grossly polluted by industrialisation, but recovering after concerted conservation and pollution control activities.

In this way, then, all of our game fishes serve as more than merely indicators of healthy waters. All of them in one way or another can be regarded as iconic of the ecosystems in which they occur. Far from being a constraint on economic progress and associated human wellbeing, these iconic species are in fact indicators of the vitality of the fundamental natural resources that make this possible.

18.6 What have our game fishes ever done for us?

In years gone by, game fishing and the maintenance of healthy stocks of salmonid fishes to support it may have been construed as solely for the diversion of privileged sectors of society. This class-ridden caricature is perhaps reinforced in Scotland, where rights to salmon fishing are related to heredity rather than licence. However, in a more egalitarian and environmentally literate age, more and more of us not only have access to game fishing, but also realise that the very existence of these fishes of the salmon family is emblematic of so many more benefits that healthy water environments bestow upon all in society. From sport to trade, food sources, fresh water, diverse wildlife and our sense of place and tranquillity, Atlantic salmon, trout, Arctic charr and grayling assure us of a place fit to support our diverse needs now and tomorrow.

In years gone by, we might have asked why we should allocate limited financial resources and supplies of water in our rivers to the needs of wildlife, when people have so many other pressing demands upon them. This same cry is indeed still heard loudly in negotiations, both here and particularly in more arid lands. But we now have the wisdom to see that fish and other wildlife thriving in our rivers do not do so at our expense, but rather that they constitute a valuable resource and an indicator of the capacity of healthy and recovering river systems to continue to support our many needs into the future.

CHAPTER 19

Game fishes for tomorrow

Atlantic salmon, brown or sea trout, Artic charr and grayling are undoubtedly under considerable threat across the British Isles. Were they merely to be incidental elements of our fauna, important only for some peripheral and altruistic enjoyment, this would be a shame but not a disaster. However, as we have seen, the fishes of the salmon family are far more significant than that. They are indeed iconic of the waters that characterise and support both their, and our, lives and lifestyles. The fate of these fishes can truly be said to be intimately interconnected with our own potential to lead healthy and fulfilled lives. That we have game fishes for tomorrow is not only important; it is essential.

19.1 Game fishes for everyone

It is not, sadly, possible to be fully egalitarian in granting access to salmon and trout angling to all of society. Would that our environment was sufficiently pristine to sustain thriving populations of naturally representative species right across the country. And, of course, we also have to contend with a human population that is not only vast and growing, but also decidedly clumped in distribution across the British Isles and Europe, as well as across much of the natural range of the world's salmonid species.

It may be doubtful whether salmon were once so abundant that we 'could walk across their backs' during spawning runs, or that the contracts of poor people included clauses precluding them from being fed salmon too often; fascinating though these myths may be, they are sadly undermined by facts discussed in Chapter 11, Net results. However, it is at least clear that our game-fish species are significantly in decline today. The pressures that we have put on the natural world throughout our pathway of industrialised development have significantly reduced populations of game and other fish species, as well as the vitality of the broader ecosystems that we depend upon to meet our needs. The prognosis of current trends for the world's freshwater fishes gives cause for serious concern and demands a proportionate response. For example, according to Kottelat and

Fig 19.1 A wild brown trout: valuable and vulnerable.

Freyhof's 2007 *Handbook of European Freshwater Fishes*, 200 of the 522 European freshwater fish species (38%) are threatened with extinction, with a further 12 already extinct, representing a much higher level of threat than that facing either Europe's birds or mammals. On a wider global scale, the 2009 IUCN 'Red List' of endangered organisms found that 1,147 of the 3,120 freshwater fish assessed (37%) are now threatened with extinction, underlining a stated view that 'Creatures living in freshwater have long been neglected'.

Estimates of the decline in the Atlantic salmon population vary, but there is unanimity that the decline is widespread, significant and showing no serious sign of sustained improvement. An accumulation of multiple pressures, significantly including changes in the marine environment leading to reduced sea survival and degradation of the freshwater and estuarine environment limiting production in home waters, has resulted in declining Atlantic salmon numbers across the North Atlantic in the last 30–40 years. The Atlantic Salmon Trust estimates that numbers of Atlantic salmon returning to spawn have dropped by over 50% since the 1970s. Furthermore, with stocks generally in decline, the return of so few spring salmon demands special conservation measures and that, in addition, sea trout are almost extinct in some rivers on the west coasts of Scotland and Ireland. The situation is already parlous indeed.

Fig 19.2 Another of Robin Armstrong's marvellous depictions of a wild brown trout in pure river water.

At the same time, the global human population of seven billion reflects a doubling of human mouths to feed and feet to tread on this planet within the lifetimes of the authors. Not only are nature's resources dwindling, but our demands upon them are spiralling. We are long overdue for a radical reform of our concept of human progress, redefining it as one that respects the resources that make progression possible, and the ecosystems

Fig 19.3 At the end of the rainbow: a brighter future for fish and people.

and iconic species that indicate whether or not we are living sustainably. Take away the salmon and we lose the reassurance that nature is working well, and for the benefit of all. Lose the trout, and there is something just 'not right', and inherently fragile, about an ecosystem robbed of one of its most iconic constituents.

Also, the distribution of game fish species is today almost inversely related to the density of human settlement across the British Isles. The reason for this is simply that people thrive where the soil is fertile and flat enough to grow food and build a sprawl of infrastructure, and the rivers are wide and languid enough to yield plentiful water for domestic and industrial use, to transport goods and dilute wastes. Species of the salmon family, conversely, thrive best where rivers flow cool and rapid, or in deep and well-oxygenated lakes, and so generally where the gradients and geology are less suited to urban and agricultural uses. Here, nutrients are naturally in short supply and habitats are scoured, not choked with silt and not sullied too badly by the detritus of 'civilisation'.

Access to angling for 'natural', self-sustaining stocks of game fishes is therefore limited by the carrying capacity of suitable environments, often not particularly close to urban centres through which fish generally pass only on migratory runs between fresh and salt water. Market forces, for better or worse, mean that access to wild game fish stocks attract a premium that may put them beyond the reach of some sectors of society.

However, as already discussed, habitats of a quality necessary to support self-sustaining wild game fish stocks have far wider benefit to society. Buffered flows of fresh water from healthy uplands nourish lowland rivers, and are abstracted into public water supplies requiring less treatment and, therefore, lower chemical and energy inputs and associated financial costs. Healthy river systems provide amenity and leisure in many different forms, as well as educational resources and places of spiritual renewal. They also support valuable and diverse wildlife, regulate flooding and air quality, recycle nutrients and renew soils. And by no means the least benefit of rivers and associated wetlands and floodplains is felt by our global society, through their ability to store carbon and, hence, help to regulate our increasingly less stable climate.

Fig 19.4 A vibrant river supports people and fish alike.

There may be an argument in some sections of society that we do not need game anglers, notwithstanding the social, economic and especially environmental benefits which accrue from angling and fisheries management. However, what cannot be argued with is that the game fish species themselves, and the environments upon which they depend, are vital for the benefit of all. If we want a decent future for ourselves and our planet, we need our game fishes and the pristine aquatic environments that support them. Their mere existence, as we have seen throughout this book, is proof that we are still doing something right with our management and conservation policies, whereas their abundance in a given river system is a vital indication that we are hitting our environmental targets. And, if we are doing that, then we are indeed meeting at least some of our responsibilities to use this planet's precious, and vitally important, natural resources sustainably, such that we pass on natural beauty, diversity and opportunity to our children and others yet to come.

Fig 19.5 A magnificent spring salmon, fresh off the tide, about to be returned to the river. Reward for all the hard fisheries management effort that nurtured this fish from egg to returning adult, and now with every chance of surviving to breed in the autumn and set the whole cycle going again.

And, of course, we have seen the elegant adaptations of Atlantic salmon and other game fishes that have enabled them to withstand climatic variability as extreme as ice ages, surviving to colonise the new waters left behind in their wake. This should provide us with at least some crumbs of hope that nature can reclaim this world and provide for us the many benefits that support our long-term well-being, if we can only learn to tread more lightly upon it.

For our game fishes, for ourselves and for the future, we have to learn, and to apply these lessons, with the utmost priority.

Bibliography and information resources

All Ireland Species Action Plans. (2009) http://www.doeni.gov.uk/niea/biodiversity/sap_uk/all_ireland_species_action_plans.htm.

Atlantic Salmon Trust. http://www.atlanticsalmontrust.org.

Bailey, J. (1994) *Salmon Fishing: In Search of Silver*. Ramsbury, The Crowood Press.

Bern Convention. (1979). The Bern Convention on the Conservation of European Wildlife and Natural Habitats. Strasbourg, Council of Europe.

Beville, S. (2005) The economic significance of the fisheries of the Test and Itchen. *The Test and Itchen Association Ltd Rivers Report*, pp.20–3. Romsey, The Test and Itchen Association Ltd.

British Trout Association. http://www.britishtrout.co.uk.

Byorndal, T., Knapp, G. A. & Lem, A. (2003) *Globefish Research Programme. Vol. 73: Salmon – A Study of Global Supply and Demand*. Portland, Oregon, Ecumenical Ministries of Oregon's Interfaith Network for Earth Concerns.

Carson, R. (1962) *Silent Spring*. London, Hamish Hamilton.

Carty, P. & Payne, S. (1998) *Angling and the Law*. Ludlow, Merlin Unwin Books.

Cefas. (2007) *Multi-species Fisheries Management: A Comprehensive Impact Assessment of the Sandeel Fishery Along the English East Coast*. CEFAS Contract report M0323/02. Lowestoft, Centre for Environment, Fisheries and Aquaculture Science.

Defra. (2009) *Public Attitudes and Behaviours Towards the Environment – Tracker Survey*. London, Department for Environment, Food and Rural Affairs. http://www.defra.gov.uk/evidence/statistics/environment/pubatt/download/report-attitudes-behaviours2009.pdf.

Diamond, J. (2004) *Collapse: How Societies Choose to Fail or Succeed*. New York, Viking.

Eden Rivers Trust. http://www.edenriverstrust.org.uk.

Environment Agency. (2008) *Better Sea Trout and Salmon Fisheries: Our Strategy for 2008–2021*. Bristol, Environment Agency. http://publications.environment-agency.gov.uk/pdf/GEHO0608BNWT-e-e.pdf?lang=_e.

Environment Agency. (2009) *Salmon Stock Assessment Report: Highlights for 2008*. Bristol, Environment Agency. http://www.environment-agency.gov.uk/research/library/publications/106767.aspx.

Environment Agency. (2009) *Fisheries Statistics Report 2008: Salmonid and Freshwater Fisheries Statistics for England and Wales, 2008 (Declared Catches of Salmon and Sea Trout by Rods, Nets and Other Instruments)*. Bristol, Environment Agency. http://www.environment-agency.gov.uk/static/documents/Business/53052_EA_Fisheries_Report_2008.pdf.

Environment Agency. (2009) *National Trout and Grayling Fisheries Strategy*. Bristol, Environment Agency. http://www.environment-agency.gov.uk/static/documents/Business/Trout_and_grayling_strategy.pdf.

Environment Agency. (2010) *Our River Habitats. The State of River Habitats across England, Wales and the Isle of Man: A Snapshot*. Bristol, Environment Agency.

European Community. (2000) Directive 2000/60/EC of the European Parliament and of the Council of 23 October 2000 establishing a framework for Community action in the field of water policy. http://eur-lex.europa.eu/LexUriServ/LexUriServ.do?uri=CELEX:32000L0060:EN:NOT.

Everard, M. (2004) Investing in sustainable catchments. *The Science of the Total Environment* **324(1–3)**:1–24.

Everard, M. (2005) *Water Meadows*. Ceredigion, Forrest Text.

Everard, M. (2008) *PVC: Reaching for Sustainability*. London, IOM3 and The Natural Step.

Everard, M. (2009) *The Business of Biodiversity*. Ashurst, WIT Press.

Everard, M. (2010) *Ecosystem Services Assessment of Sea Trout Restoration Work on the River Glaven, North Norfolk*. Environment Agency Evidence Report SCHO0110BRTZ-e-e. Bristol, Environment Agency.

Everard, M. (2011) *Common Ground: The Sharing of Land and Landscapes for Sustainability*. London, Zed Books, 214pp.

Everard, M. (in press) *Freshwater Fishery Ecosystems and their Services to Society*. Proceedings of the 42nd IFM Conference 'The Rejuvenating Role of Urban Fisheries in the Big Society', Oxford, 18–20 October 2011.

Everard, M. & Appleby, T. (2009) Safeguarding the societal value of land. *Environmental Law and Management* **21**:16–23.

Everard, M. & Jevons, S. (2010) *Ecosystem Services Assessment of Buffer Zone Installation on the Upper Bristol Avon, Wiltshire*. Environment Agency Evidence Report SCHO0210BRXW-e-e. Bristol, Environment Agency.

FAO. (2000) *The State of World Fisheries and Aquaculture 2000*. Report. Rome, United Nations Food and Agriculture Organization. http://www.fao.org/docrep/003/X8002E/x8002e00.HTM.

FAO. (2005) *Yearbook of Fisheries Statistics Extracted with Fishstat Version 2.30*. Fisheries database: aquaculture production quantities 1950–2003; aquaculture production values 1984–2003; capture production 1960–2003; commodities production and trade 1976–2002. Rome, United Nations Food and Agriculture Organization. http://www.fao.org/fi/statist/FISOFT/FISHPLUS/asp.

Fisheries Research Services. (2003) *Sandeels in the North Sea*. Paper No.ME01A 03 03. Edinburgh, Fisheries Research Services (an agency of the Scottish Executive). http://www.frs-scotland.gov.uk.

Forest Stewardship Council. http://www.fsc.org.

GHK Consulting Ltd. (2004) *Revealing the Value of the Natural Environment in England*. A report to the Department for Environment, Food and Rural Affairs, March 2004. Plymouth, GHK Consulting Ltd. https://statistics.defra.gov.uk/esg/reports/rvne.pdf.

Giles, N. (1994) *Freshwater Fish of the British Isles*. Shrewsbury, Swan Hill Press, 192pp.

Giles, N. (2005) *The Nature of Trout*. Verwood, Perca Press, 233pp.

Giraldus Cambrensis. (1982) *The History and Topography of Ireland*. Translated by J. J. O'Meara (1951). Harmondsworth, Penguin Books, 57pp.

Habitats Directive. (1992). Council Directive 92/43/EEC of 21 May 1992 on the conservation of natural habitats and of wild fauna and flora. Strasbourg, Council of Europe.

Hardy and Greys. http://www.hardyfishing.com.

Houghton, Rev. W. (1879) *British Fresh-water Fishes*. [Various impressions are available, e.g. 1981, Cullompton, Web & Bower (Publishers) Ltd.]

Hunter Committee. (1963) Also known as the Scottish Salmon and Trout Committee. Edinburgh, Department of Agriculture & Fisheries (Scotland).

Intergovernmental Panel on Climate Change. http://www.ipcc.ch.

International Union for the Conservation of Nature (IUCN). 'Red List' of endangered organisms. http://www.iucnredlist.org.

Johnston, G. (2002) *Arctic Charr Aquaculture*. Oxford, Fishing News Books.

Joint Nature Conservation Committee. (2007) *Second Report by the UK under Article 17 on the Implementation of the Habitats Directive from January 2001 to December 2006*. Peterborough, JNCC. http://www.jncc.gov.uk/article17.

Kearns, J. P. (2005) Current production methods for the production of salmon feeds. *International Aquafeed* **8(1)**:28–33.

Kottelat, M. & Freyhof, J. (2007) *Handbook of European Freshwater Fishes*. Cornol, Switzerland (Kottelat) and Berlin, Germany (Freyhof), privately published by the authors. ISBN 978-2-8399-0298-4, xiv+646pp.

Le Cren, D. (2001) The Windermere perch and pike project: a historical review. *Freshwater Forum* **15**:3–34.

Leopold, A. (1949) *A Sand County Almanac: And Essays on Conservation from Round River*. New York, Oxford University Press.

Maloti Drakensberg Transfrontier Project. (2007) *Payment for Ecosystem Services: Developing an Ecosystem Services Trading Model for the Mnweni/Cathedral Peak and Eastern Cape Drakensberg Areas*. Mander, M. ed. INR Report IR281. Development Bank of Southern Africa, Department of Water Affairs and Forestry, Department of Environment Affairs and Tourism, Ezemvelo KZN Wildlife, South Africa.

Marine Stewardship Council. http://www.msc.org.

Marine Stewardship Council. (2009) *Net Benefits: The First Ten Years of MSC Certified Sustainable Fisheries*. London, Marine Stewardship Council. http://www.msc.org/documents/fisheries-factsheets/net-benefits-report.

Marine Stewardship Council. (2011) *Annual Report 2010/11*. London, Marine Stewardship Council. http://www.msc.org/documents/msc-brochures/annual-report-archive/annual-report-2010-11-english.

Millennium Ecosystem Assessment. (2005) *Ecosystems & Human Well-being: Synthesis*. Washington, DC, Island Press.

North Atlantic Salmon Conservation Organization. http://www.nasco.int.

North Atlantic Salmon Fund. http://www.nasfworldwide.com.

Orvis. http://www.orvis.co.uk.

Palmer, M. & Finlay, V. (2003) *Faith in Conservation: New Approaches to Religions and the Environment*. Washington, DC, The World Bank.

Perrot-Maître, D. (2006) *The Vittel Payments for Ecosystem Services: A 'Perfect' PES Case?* London, International Institute for Environment and Development.

Plunket Greene, H. (1924) *Where the Bright Waters Meet*. [Republished in 2007 by the Medlar Press, Ellesmere, UK.]

Ravenga, C., Murray, S., Abramovitz, J. & Hammond, A. (1998) *Watersheds of the World: Ecological Value and Vulnerability*. Washington, DC, The World Resources Institute and the Worldwatch Institute.

Reynolds, J. D. (1988) *Ireland's Freshwaters*. Dublin, The Marine Institute.

Salmon & Trout Association. http://www.salmon-trout.org.

Salmon Aquaculture Dialogue. http://www.worldwildlife.org/salmondialogue.

SCaMP. http://www.unitedutilities.com/?OBH=3128.

Shelton, R. (2009) *To Sea and Back: The Heroic Life of the Atlantic Salmon*. London, Atlantic Books.

Stern, N. (2006) *The Economics of Climate Change*. London, HM Treasury. http://www.hm-treasury.gov.uk/sternreview_index.htm.

Thames Rivers Restoration Trust. http://www.trrt.org.uk.

The Rivers Trust. www.theriverstrust.org.

Tweed Foundation. http://www.tweedfoundation.org.uk.

UK Biodiversity Action Plan. http://jncc.defra.gov.uk/default.aspx?page=5155.

UNDP Human Development Reports. http://hdr.undp.org/en/reports/global/hdr2009/.

United Nations Development Programme (UNDP). http://www.undp.org.

Walton, I. (1653) *The Compleat Angler*. [Available in many impressions.]

Westcountry Rivers Trust. http://www.wrt.org.uk.

Wheeler, A. (1969) *The Fishes of the British Isles and North West Europe*. East Lansing, Michigan State University Press, 613pp.

Wild Trout Trust. http://www.wildtrout.org.

Wildlife and Countryside Act. (1981) http://www.legislation.gov.uk/ukpga/1981/69.

World Commission on Dams. (2000) *Dams and Development: A New Framework for Better Decision-making*. London, Earthscan.

Wye and Usk Foundation. http://www.wyeuskfoundation.org.

Index

Note: Page numbers in *italic* refer to figures.